# Lecture Notes in Mathematics

Edited by A. Dold and B. Eckmann

Subseries: Nankai Institute of Mathematics, (Tianjin, P.R. China) vol. 4

Adviser: S.S. Chern

1336

Bernard Helffer

# Semi-Classical Analysis for the Schrödinger Operator and Applications

Springer-Verlag

Berlin Heidelberg New York London Paris Tokyo

**Author**

Bernard Helffer
Département de Mathématiques
U.A.CNRS 758, Université de Nantes
2, Rue de la Houssinière, 44072 Nantes Cédex 03, France

Mathematics Subject Classification (1980): 35P20; 58C40; 81C05; 81C10; 81C12

ISBN 3-540-50076-6 Springer-Verlag Berlin Heidelberg New York
ISBN 0-387-50076-6 Springer-Verlag New York Berlin Heidelberg


Printed in Germany

Printing and binding: Druckhaus Beltz, Hemsbach/Bergstr.
2146/3140-543210

## §0 INTRODUCTION.

This course falls into two different parts. The first part (Sections 1-5) is the written version of ten lectures I gave in Nankai University in October 1985. It can be seen as an introduction to my work with J. Sjöstrand ($[HE\text{-}SJ]_{1-6}$). My purpose was to give in a simpler situation a relatively self-contained presentation of the tunneling effect. In fact, we have tried to refer only to the two basic books of Reed-Simon [RE-SI] and Abraham-Marsden. [AB-MA] (see also Abraham-Robin for the theory of the stable manifolds). The material presented here comes essentially from $[HE\text{-}SJ]_1$ but we have also used improvements that we found later in $[HE\text{-}SJ]_{2-6}$ and the proof presented here is, at least in the form, partly different (particularly for the B.K.W construction, where we present a simpler method, less general, but perhaps easier to understand for non-specialists in microlocal analysis).

Almost two years later, in June 1987, I was asked to complete these notes to permit a publication as a volume of the Springer Lecture Notes (Nankai Subseries). During these two years, we had applied these techniques, in collaboration with J. Sjöstrand or through students, to many other problems where the tunneling effect plays an important role: resonances, Schrödinger with periodic potential, Schrödinger with magnetic fields, etc... but it is probably too early to write a definite book on the subject. At the same time, a very good book on the Schrödinger operator by Cycon-Froese-Kirsch-Simon [C.F.K.S] has appeared. We have therefore chosen to present in § 6 and § 7 subjects which are complementary to this book and which are natural applications of the theory developed in the first 5 sections.
This book is organized as follows.

In § 1, we present a brief survey of semi-classical mechanics and recall basic facts on the Schrödinger operator. This material is more developed in the recent book by D. Robert [Ro] which we recommend to the reader interested in pseudodifferential techniques.
§ 2 is concerned with the B.K.W construction at the bottom. In § 3, we study the decay of the eigenfunctions in the spirit of Agmon [AG]. These results were developed in the semi-classical context by B. Simon $[SI]_{2-4}$ and B. Helffer-J. Sjöstrand $[HE\text{-}SJ]_{1-9}$.
§ 4 is concerned with the interaction between different wells. This is a mathematical version of the well-known L.C.A.O method used by chemists.
In § 5, we present briefly the application to Witten's proof of the Morse inequalities [WIT]. There is an intersection with one chapter of the book [C.F.K.S] but we have tried to go a little further using the techniques of sections 2,3,4, however we still remain far from the best results (related to the method of instantons) obtained in $[HE\text{-}SJ]_4$. In § 6, we study the asymptotic behavior of the first band of

the Schrödinger operator with periodic potentials and present results obtained by B.Simon $[SI]_6$ and A. Outassourt [OU].

§ 7 is devoted to the study of some classical problems on the Schrödinger operator with magnetic fields: criteria for the compactness of the resolvent (after Helffer-Mohamed) [HE-MO] ), multiplicity of the first eigenvalue (after Avron-Herbst-Simon [A-H-S], Lavine-O'Carroll [LA, O'CA], Helffer-Sjöstrand $[HE\text{-}SJ]_{10}$ and Helffer [HE]), effect of the flux of the magnetic field $[HE]_3$). The study of these problems is only beginning and we just give a flavor of some of the problems (see also the chapter in [C.F.K.S] devoted to these questions).

I have many people to thank at the end of this introduction. First of all the Universities of Wuhan and Nankai which organized this course in October 1985 with the help of the French "Ministère des Relations Extérieures" and particularly Professors Chi Min-yu, Wang Rou-hai and S.S. Chern.
I want also to thank J. Sjöstrand and D. Robert with whom I have collaborated in this field, M. Dauge who read a part of the text and C. Brunet and M. Coignac who typed the manuscript.
For the reader who does not appreciate my poor English, let me mention in closing that there exists a Chinese version of this course, written up by Professor Chi Min-yu and his students.

# CONTENTS

## §1 GENERALITIES ON SEMI-CLASSICAL ANALYSIS.

The purpose of the semi-classical analysis is to understand, from a mathematical point of view, the general correspondance principle of the Quantum mechanics saying that, when the Planck constant h tends to zero, we must recover, starting from the Quantum mechanics, the classical mechanics. The best references for this section are for example : Fedoryuk-Maslov [FE-MA], D. Robert [RO], J. Leray [LE] for the semi-classical aspects, Arnold [AR] and Abraham-Marsden [AB-MA] for the classical mechanics and Reed-Simon [RE-SI] for the study of the Schrödinger Operator.

### §1.1 - The Classical mechanics (See section 3.3 in [AB-MA]).

In the most simple cases, the classical mechanics describes the motion of a point x(t) in a space $\mathbb{R}^n_x$ (x is the position) and more generally in a $C^\infty$ manifold M. But adding the impulsion $\xi(t)$ of the point , we prefer to work in $\mathbb{R}^n_x \times \mathbb{R}^n_\xi$ and more intrinsically in $T^*M$ the cotangent bundle to M. $T^*M$ is a symplectic manifold, that means that we have on $T^*M$ a canonical non-degenerate, closed 2 -form $\omega$. In the case of $T^*\mathbb{R}^n$ , $\omega$ is defined by :

$$(1.1.1)\qquad \omega = \sum_j d\,\xi_j \wedge d\,x_j$$

In the case of $T^*M$ , if $(x_j)$ is a system of local coordinates and $(\xi_j)$ is the dual system of coordinates, $\omega$ can be written in the same way.

More generally, if $\Omega$ is a symplectic manifold of dimension 2n, we can always find locally a system of coordinates $(x,\xi)$ s.t $\omega$ is defined by (1.1.1). These coordinates are called the canonical coordinates.

In Hamiltonian Mechanics, the motion is described by a $C^\infty$ function on $T^*M = \Omega$ called the hamiltonian :

$$(1.1.2)\qquad (x,\xi) \to p(x,\xi)$$

Associated to this hamiltonian, we define the hamiltonian vector $H_p$ on $\Omega$ , which is given in canonical coordinates by :

$$(1.1.3)\qquad H_p = \left(\frac{\partial p}{\partial \xi}\, , -\frac{\partial p}{\partial x}\right)$$

The motion of a point in $\Omega$ is described by the integral curves of $H_p$ (called the bicharacteristics) which are the solutions of the system :

$$(1.1.4)\qquad \begin{cases} \dfrac{dx}{dt} = \dfrac{\partial p}{\partial \xi}(x,\xi) & x(0,y,\eta) = y \\[2ex] \dfrac{d\xi}{dt} = -\dfrac{\partial p}{\partial x}(x,\xi) & \xi(0,y,\eta) = \eta \end{cases}$$

It is not our purpose to give here precise existence theorems for the equations (1.1.4), but it is well-known that at least locally and for $|t|$ small, the solutions exist and we can define the Hamiltonian flow $\phi_t$, by :

(1.1.5) $$\phi_t(y,\eta) = (x(t,y,\eta),\xi(t,y,\eta))$$

We are mainly interested in these lectures by the Hamiltonian :

$$p(x,\xi) = \xi^2 + V(x)$$

In the case where M is a Riemannian manifold, if $G = g^{ij}$ is the matrix of the metric in the coordinates x, we define $G^{-1} = g_{ij}$ and we must define $\xi^2$ as :

(1.1.6) $$\xi^2 \overset{\text{def}}{\equiv} \sum_{i,j} g_{ij}(x)\, \xi_i\, \xi_j$$

--

Then the motion is given (in the case of $\mathbb{R}^n$) by :

(1.1.7) $$\frac{dx}{dt} = 2\xi \ , \quad \frac{d\xi}{dt} = -\frac{\partial V}{\partial x}$$

so we recover the classical equation of the motion in M :

(1.1.8) $$\frac{d^2x}{dt^2} = -2\frac{\partial V}{\partial x} = -2 \text{ grad } V$$

(the number 2 appears because usually one takes $p(x,\xi) = \frac{\xi^2}{2} + V(x)$)

## §1.2 - The Quantum Mechanics

One of the problems is to find a natural Hilbert space. Here, because, we consider only the case when $\Omega = T^* M$, the natural choice is $L^2(M)$ (where the measure is the canonical measure associated to the Riemannian Structure ; because we consider only the case $M = \mathbb{R}^n$ or the case M = compact $C^\infty$ Manifold, $L^2(M)$ is complete). We need also a dense subspace (usually $C^\infty(M)$ if M is compact and $C_0^\infty(\mathbb{R}^n)$ or $\mathcal{S}(\mathbb{R}^n)$ in the case of $\mathbb{R}^n$). Let us consider the case of $\mathbb{R}^n$. Under some conditions on p (See [HO], the Weyl-Calculus, for a general point of view), we associate to the Hamiltonian p an operator a priori defined on $\mathcal{S}(\mathbb{R}^n)$ by the so-called Weyl Quantification :

$$p \to Op^W(p) = p^W(x,h\,D_x)$$

(1.2.1) $$\begin{cases} Op^W(p).f = h^{-n} \int\int e^{\frac{i}{h}\langle x-y,\xi\rangle}\, p(\frac{x+y}{2},\xi)\, f(y)\, dy\, đ\xi \\ \text{for } f \in \mathcal{S}(\mathbb{R}^n),\ h \in\, ]0,h_0] \\ \text{with the convention } đ\xi = (2\pi)^{-n}\, d\xi \end{cases}$$

via a theory of pseudodifferential operators on $\mathbb{R}^n$ .

This choice is not the only possible but it is very convenient because you get for example that, if p is real, $Op^W(p)$ is formally self-adjoint, that means :

(1.2.2) $$(p^W(x,hD)\,u/v) = (u|\,p^W(x,hD)v) \qquad \forall u \in \mathcal{S}(\mathbb{R}^n),\ \forall v \in \mathcal{S}(\mathbb{R}^n)$$

where ( / ) denotes the $L^2$ scalar product.
For our purpose, we are interested in the extension of this formally self-adjoint operator defined on $\mathcal{S}(\mathbb{R}^n)$ as a self-adjoint non-bounded operator on $L^2(\mathbb{R}^n)$. When this extension exists and is unique, $p^W(x,hD)$ is called essentially self-adjoint. General criteria to verify that $p^W(x,hD)$ is essentially self-adjoint for h small enough are given in the more general context of the " admissible operators " (associated to hamiltonians depending on h

$$p(x,\xi,h) \sim \sum_{j=0}^{\infty} h^j\, p_j(x,\xi))$$

are given in $[RO]_3$ and $[HE\text{-}RO]_{1\ \text{to}\ 3}$.
In the case of a Riemannian Manifold, the h-pseudodifferential calculus can also be defined but you lose the notion of the Weyl-Calculus at least if you don't add a group structure (See [MELI]). But, in these lectures, we will be mainly interested in the study of the Schrödinger operator associated to the Hamiltonian $\xi^2 + V(x)$, where V is a $C^\infty$ real function. In this particular case, there is a natural quantification given by the geometry. We associate to $\xi^2$ the Laplace-Beltrami operator on the Manifold M $h^2(-\Delta) = h^2(d^*+d)^2$, which is given in coordinates by :

(1.2.3) $$-h^2\,\Delta = -h^2 \sum_{i,j} \sqrt{g}\,\frac{\partial}{\partial x_i}\,\frac{1}{\sqrt{g}}\cdot g_{ij}\,\frac{\partial}{\partial x_j}$$

where $g = (\det G)$
and we get the Schrödinger operator :

(1.2.4) $$-h^2\,\Delta + V$$

which is formally self-adjoint on $C_0^\infty(M)$ where the scalar product is given by $(u,v) \to \int u\,.\,\bar{v}\,g^{-1/2}\,dx$ for u and v with support in the chart.

## §1.3 - Some basic results on the Schrödinger operators

In the case where M is a $C^\infty$ compact riemannian manifold, the Schrödinger operator $(-h^2\,\Delta + V)$ defined on $C^\infty(M)$ by (1.2.3) has a unique self-adjoint extension whose domain is the Sobolev space $H^2(M)$. Then we know that the injection of $H^2(M)$ in $L^2(M)$ is compact. In this case, it is well-known that we have a orthonormal basis of $L^2(M)$ constituted by eigenfunctions (in $C^\infty(M)$) $\varphi_j(h)$ . ($j \in \mathbb{N}$) satisfying to :

$$(1.3.1)\quad \begin{cases} (-h^2\,\Delta + V)\,\varphi_j(h) = \lambda_j(h)\,\varphi_j(h) \\ \lambda_j(h) \leqslant \lambda_{j+1}(h) \end{cases}$$

Moreover, for h fixed, $\lambda_j(h) \xrightarrow[j\to\infty]{} \infty$

The spectrum of the Schrödinger operator on a non compact manifold is more complicate. We will restrict ourselves in these notes in the case of $\mathbb{R}^n$ and we assume in this case that the following hypothesis is satisfied for some constant $C_0$ :

(1.3.2) $\quad V \in C^\infty(\mathbb{R}^n)\,,\quad V \geqslant -C_0$

Under this hypothesis, one can define a self-adjoint operator on $L^2(\mathbb{R}^n)$ by taking the Friedrichs extension starting of $C_0^\infty(\mathbb{R}^n)$. Moreover this is the unique self-adjoint extension of $-h^2\,\Delta + V$ starting of $C_0^\infty(\mathbb{R}^n)$ (See [RE-SI] Vol. II, Th. X.28 and p. 340 ex 24). Let us define :

(1.3.3) $\quad C = \underline{\lim}_{|x|\to\infty}\; V$

Then, the restriction of the spectrum to $]-\infty, C[$ is constituted of eigenvalues with finite multiplicity (See Reed-Simon[RE-SI], § XIII.4 Cor. 2 p. 113 and Th. XIII.16).

In the future, we make the convention that $C = +\infty$ in the compact case.

As an example of a semi-classical result we shall need after, let us present a spectral result.

For $\lambda < C$, let us define :

(1.3.4) $\quad N_h(\lambda) = \{\#\ j,\ \lambda_j(h) \leqslant \lambda\}$

The problem is to study the asymptotic behavior of $N_h(\lambda)$ when h tends to 0. It is a classical result that :

(1.3.5) $\quad \lim_{h\to 0} h^n\, N_h(\lambda) = \int_{\xi^2+V(x)<\lambda} dx\, d\!\!{}^-\!\xi$

(See [RE-SI], [HE-RO], [CDV] for references).

In the last years, many mathematicians have tried to give the best estimate for :

(1.3.6) $\quad R_h(\lambda) = h^n\, N_h(\lambda) - \int_{\xi^2+V(x)<\lambda} dx\, d\!\!{}^-\!\xi$

Let us give the best known theorem

Theorem 1.3.1

Let V verifying (1.3.2).
Suppose that $\lambda < C$ and that $\lambda$ is not a critical value for V then :

$$R_h(\lambda) = 0(h) \tag{1.3.7}$$

This theorem is proved in the compact case by Colin de Verdière [C.D.V]$_1$. In the case of $\mathbb{R}^n$, $R_h(\lambda) = 0(h^{\delta})$ was proved by Tulovski-Šubin [ŠU] (for $\delta < 1/2$), Hörmander [HO]$_2$ ($\delta < 2/3$), Combes-Schrader-Seiler [C.S.S], and Helffer-Robert [HE-RO]$_3$ but always under additionnal hypothesis of the potential of the following type :

$$\left\{ \begin{array}{lll} \exists\, C_\alpha & \text{s.t} & |\partial_x^\alpha V| \leq C_\alpha (V + C_0 + 1) \\ \exists\, D,M & \text{s.t} & |V(x)| \leq D(V(y) + C_0 + 1)\ (1 + |x-y|)^M \\ & & \forall x \in \mathbb{R}^n,\ \forall y \in \mathbb{R}^n . \end{array} \right. \tag{1.3.8}$$

Ivrii has announced theorem 1.3.1. without hypothesis (1.3.8). We shall explain in section 4.2, Remark 4.2.4. how to deduce the Theorem 1.3.1. in the general case from the Theorem (1.3.1) in the particular casewhere (1.3.8) is satisfied.

Remark 1.3.2

In the following we need only a weak version of (1.3.5) :

$$N_h(\lambda) = \mathcal{O}(h^{-N_0}) \qquad \text{for some } N_0. \tag{1.3.9}$$

but also in the study of Dirichlet problems.
This type of results is an easy consequence of the min-max principle ([RE-SI] Vol. IV, Th. XIII.1 and XIII.2 and probl. 1 p. 364) which gives, if $\lambda \leq \underline{\lim}_{|x| \to \infty} V_0$, the inequality :

$$N_h^{V_0}(\lambda) \geq N_h^{V_1}(\lambda) \quad \text{if} \quad V_0 \leq V_1 \tag{1.3.10}$$

Remark 1.3.3

One could think that it is rather stupid to improve the estimate of $R_h(\lambda)$.
But O(h) is an important step because this estimate cannot be improved without adding extra-hypotheses on the flow $\phi_t$ on the energy level : $\xi^2 + V = \lambda$ . Ivrii [IV] and V. Petkov--D. Robert [PE-RO] have given conditions to get 0(h).

Remark 1.3.4

Interesting questions remain, in the case of Magnetic fields, that means for operators of the type :

$$P(h) = - \sum_{j=1}^{n} (h\,\partial_{x_j} - i\,a_j)^2 + V$$

We refer to papers of Avron-Herbst-Simon [A.H.S], J.M. Combes - R. Schrader - R. Seiler [C.S.S], D. Robert $[RO]_2$ , B. Simon $[SI]_1$ and more recently to three papers of H. Tamura [TA], J.P. Demailly [DE] and Y. Colin de Verdière $[C.D.V]_2$.

Another interesting point is that you can get, under convenient hypotheses, a compact resolvant without the hypotheses that $V \to \infty$.

For example, one can deduce from my results with Nourrigat [HE-NO] that, if

$$V = \sum_{i=1}^{N} q_i^2(x)$$

and if the vector space of the polynomials generated by the $\partial_x^\alpha q_i$ contains all the polynomials of d° 1, the resolvant is compact.

Example : $V = x_1^2\, x_2^2$ on $\mathbb{R}^2_{x_1,x_2}$

The estimate of $N_h(\lambda)$ (for $\lambda \to \infty$ (h fixed) or $h \to 0$ ($\lambda$ fixed)) is a difficult task in general (See for particular cases $[RO]_2$ , $[SI]_6$ ).

Another interesting example is :

$$-(h\,\partial_{x_1} - x_2^2)^2 - (h\,\partial_{x_2})^2 + x_1^2$$

which is also with compact resolvant.

Remark 1.3.5.

There are many papers studying the asymptotic behavior of $N_h(\lambda)$ when $\lambda$ tends to $\infty$. I prefer to refer to my book [HE] where many references are given.

## §2 B.K.W CONSTRUCTION FOR A POTENTIAL NEAR THE BOTTOM IN THE CASE OF NON-DEGENERATE MINIMA.

In all this section, we work with some $C^\infty$ potential V which admits a local non-degenerate minimum at a point.
By changing of coordinates, we can suppose that :

$$V(0) = 0,\ V'(0) = 0,\ V''(0) > 0 \tag{2.0}$$

where $V''(0)$ is the Hessian of V at 0.
In this section we will not try to follow the most direct way to get the results but we prefer to see how the different technics work.

### §.2.1 - The Harmonic oscillator

Before to study the general situation, it is convenient to recall the basic properties of the Harmonic oscillator.
Let us consider in $\mathbb{R}^n$ :

$$P_0(h) = - \sum_{j=1}^{n} h^2 \frac{d^2}{dx_j^2} + \sum_{j=1}^{n} \mu_j x_j^2 \qquad \text{with } \mu_j > 0 \tag{2.1.1}$$

This is of course a Schrödinger operator whose potential is given by :

$$V_0(x) = \sum_j \mu_j x_j^2 \tag{2.1.2}$$

Let us recall very briefly how to compute the spectrum and the eigenfunctions of $P_0$ (h).

Step 1 The spectrum of $(- \frac{d^2}{dx^2} + x^2)$ is constituted by eigenvalues $(2j+1)$ $(j \in \mathbb{Z}^+)$.
The first eigenfunction is given by $(\sqrt{2\Pi})^{-1}\ e^{-x^2/2} = u_0(x)$.
The $(j+1)$th eigenfunction $u_j$ corresponding to the eigenvalue $(2j+1)$ is deduced from $u_{j-1}$ by the relation :

$$u_j(x) = \alpha_j \cdot (- \frac{d}{dx} + x)\, u_{j-1} \qquad \text{where } \alpha_j > 0 \tag{2.1.3}$$

and is chosen to normalize $u_j$.
You get easily that :

$$u_j(x) = P_j(x) \cdot e^{-x^2/2} \tag{2.1.4}$$

where $P_j(x)$ is a polynomial of order j

Step 2 By easy manipulations, you get that the spectrum of $P_0(1)$ is given by

(2.1.5) $$\lambda_\alpha = \sum_{j=1}^{n} \sqrt{\mu_j}\ (2\alpha_j + 1) \quad , \quad \alpha \in (\mathbb{Z}^+)^n$$

with corresponding eigenfunction

(2.1.6) $$u_\alpha(x) = P_{\alpha_1}(\mu_1^{1/4} x_1) \cdot \dots \cdot P_{\alpha_n}(\mu_n^{1/4} x_n) \cdot e^{-\sum_{j=1}^{n} \sqrt{\mu_j}\ x_j^2/2} \cdot (\mu_1 \cdot \dots \cdot \mu_n)^{-1/8}$$

Step 3 We observe now that if $\Psi(x)$ is a normalized eigenfunction for $P_0(1)$ associated to the eigenvalues $\lambda$, $\Psi(x/\sqrt{h}) \cdot h^{-n/4}$ is a normalized eigenfunction for $P_0(h)$ associated to the eigenvalue $\lambda h$. We then get finally that the eigenvalues of $P_0(h)$ are given by :

(2.1.7) $$\lambda_\alpha(h) = h\,\lambda_\alpha(1) = \Big(\sum_{j=1}^{n} \sqrt{\mu_j}\ (2\alpha_j+1)\Big)\, h \qquad (\alpha \in (\mathbb{Z}^+)^n)$$

with corresponding eigenfunction :

(2.1.8) $$u_\alpha(h)(x) = h^{-n/4}\, u_\alpha(x/\sqrt{h}) = h^{-n/4}\, (\mu_1 \dots \mu_n)^{-1/8}\, P^\alpha\Big(\frac{\mu^{1/4}.x}{\sqrt{h}}\Big) \cdot e^{-\sum_{j=1}^{n} \sqrt{\mu_j}\, x_j^2/2h}$$

--

We keep from (2.1.8) that $u_\alpha(h)(x)$ has the following form :

(2.1.9) $$u_\alpha(h)(x) = h^{-n/4}\, a_\alpha(x,h)\, e^{-\varphi_0(x)/h}$$

where

(2.1.10) $$\varphi_0(x) = \sum_{j=1}^{n} \frac{\sqrt{\mu_j}\ .\ x_j^2}{2}$$

(2.1.11) $$a_\alpha(x,h) = c_\alpha\, h^{-|\alpha|/2}\, \Big[x^\alpha + \sum_{\substack{|\beta| < |\alpha| \\ |\beta| + \gamma/2 = |\alpha|}} c_{\alpha\gamma}^{\beta}\, x^\beta\, h^\gamma\Big] \qquad \text{with } c_\alpha \neq 0$$

Let us also remark that $\varphi_0(x)$ satisfies to the " eiconal " equation :

(2.1.12) $$|\nabla \varphi_0(x)|^2 = V_0(x)$$

§.2.2 - Approximate solutions starting from the Harmonic oscillator.

For a more sophisticated version in this spirit, we refer to B. Simon $[SI]_2$ (and to his references). We make the hypothesis (2.0), and suppose moreover that :

(2.2.1) $$V''(0) = 2\begin{pmatrix} \mu_1 & & 0 \\ & \mu_2 & \\ 0 & & \ddots\ \mu_n \end{pmatrix} \quad ; \quad \mu_j > 0$$

and we suppose (See 1.2.3) that the metric is chosen s.t :

(2.2.2) $$g = 1 \,, \quad g_{ij}(x) = \delta_{ij} + O(|x|)$$

To simplify, we shall only look at the first eigenvalue of the Schrödinger operator : $P(h) = -h^2 \Delta + V$ (see 1.2.3).
Then starting from :

(2.2.3) $$u_0(h)(x) = c_0 . h^{-n/4} e^{-\varphi_0(x)/h} \qquad (c_0 \neq 0)$$
$$\text{s.t } \|u_0(h)\| = 1$$

we introduce, for $\varepsilon > 0$ fixed sufficiently small :

(2.2.4) $$\Psi_0(h)(x) = \chi_\varepsilon(\varphi_0(x)) \,.\, u_0(h)(x)$$

where $\chi_\varepsilon$ is a $C^\infty$ function s.t $\chi_\varepsilon(t) = 1$ for $t \in [-\varepsilon/2, +\varepsilon/2]$, and supp $\chi_\varepsilon \subset [-\varepsilon,\varepsilon]$.
Let us look now at :

$$P(h)(\Psi_0(h)(x))$$

We have :

(2.2.5) $$P(h)(\Psi_0(h)(x)) = \lambda_0(h)\, \Psi_0(h)(x) + R_0(h)(x) + R_1(h)(x)$$

with

(2.2.6) $$R_0(h)(x) = [P,\chi_\varepsilon]\,(u_0(h)(x)$$

(2.2.7) $$R_1(h)(x) = \chi_\varepsilon(P-P_0(h))(u_0(h)(x))$$

It is immediate to see that :

(2.2.8) $$\|R_0(h)(x)\|_{L^2} \leqslant C\, h^{-n/4} \,.\, e^{-\varepsilon/4h}$$

Let us look now at $R_1(h)(x)$ which is the sum of two terms :

$$R^0{}_1(h)(x) = -h^2 \chi_\varepsilon \Big(\sum_{i,j} \frac{d}{dx_i} (g_{ij}(x) - g_{ij}(0)) \frac{d}{dx_j}\Big)\, u_0(h)(x)$$

and

$$R^1{}_1(h)(x) = \chi_\varepsilon(\varphi_0(x))\, (V(x) - V_0(x))\, u_0(h)(x)$$

Using (2.2.2) and the property $(V(x)-V_0(x)) = 0(|x|^3)$ we have to estimate in the $L^2$-norm

$$(a) = h^2\, 0(|x|)\, \left(\frac{d^2}{dx_i\, dx_j} u_0(h)(x)\right) . \chi_\varepsilon(\varphi_0^{\cdot}(x))$$

$$(b) = h^2 \left(\frac{d}{dx_i} u_0(h)(x)\right) . \chi_\varepsilon(\varphi_0(x))$$

and

$$(c) = 0(|x|^3) . u_0(h)(x) . \chi_\varepsilon(\varphi_0(x))$$

$$\|(a)\|^2_{L^2} \leqslant C\, h^{-n/2} \int_{\mathbb{R}^n} |x|^6 . e^{-2\varphi_0(x)/h}\, dx + C\, h^{2-n/2} \int_{\mathbb{R}^n} |x|^2\, e^{-2\varphi_0(x)/h}\, dx$$

$$\|(b)\|^2_{L^2} \leqslant C\, h^{-n/2} . h^2 \int_{\mathbb{R}^n} |x|^2\, e^{-2\varphi_0(x)/h}\, dx$$

$$\|(c)\|^2_{L^2} \leqslant C\, h^{-n/2} \int_{\mathbb{R}^n} |x|^6\, e^{-2\varphi_0(x)/h}\, dx$$

and we get easily that :

$$\|(a)\|^2 + \|(b)\|^2 + \|(c)\|^2 \leqslant \tilde{C}\, h^3$$

(2.2.9) $\quad \|R_1(h)(x)\|_{L^2} \leqslant \tilde{C}\, h^{3/2}$

We remark also very easily that :

(2.2.10) $\quad \|\Psi_0(h)(x)\|_{L^2} = 1 + \mathcal{O}(e^{-\varepsilon/4h})$

Finally, we get a $C^\infty$ function $\Psi_0(h)(x)$ s.t :

(2.2.11) $\quad (P(h) - \lambda_0(h))\, \Psi_0(x,h) = \mathcal{O}(h^{3/2})\, \|\Psi_0\| \qquad$ in $L^2(\mathbb{R}^n)$

Suppose now that

(2.2.12) $\quad \varliminf_{|x| \to +\infty} V > 0$

then using a well known property of the self-adjoint operators :

(2.2.13) $\quad d(\lambda, \mathrm{Sp}\ P(h))\, \|u\|_{L^2} \leqslant \|(P(h)-\lambda)u\|_{L^2}$

for $\lambda \in \mathbb{C}$ and where Sp P(h) is the spectrum of P(h), we get from (2.2.11) that :

(2.2.14) $\quad d(\lambda_0(h), \mathrm{Sp}\ P(h)) \leqslant \tilde{C} . h^{3/2}$

and we have proved the following proposition :

Proposition 2.2.1.

Under hypotheses (2.2.1) and (2.2.12), $P(h)$ admits at least one eigenvalue $\lambda(h)$ of $P(h)$ in the ball $B(\lambda_0(h), \tilde{C}\, h^{3/2})$ where $\tilde{C}$ is large enough, for each $h \in ]0,h_0]$ ($h_0 > 0$ sufficiently small) where $\lambda_0(h) - E_1\, h$ (with $E_1 = \sum_{j=1}^{n} \sqrt{\mu_j}$ ) is the first eigenvalue of the associated harmonic oscillator $P_0(h)$.

Remark 2.2.2.

This proposition does'nt answer to all the questions. This technique gives only the existence of some eigenvalue. The same technique gives also the existence of one eigenvalue for the Dirichlet problem in some bounded open set containing 0. Under additional hypotheses, we will prove in another section that there is only one eigenvalue in the interval $I(h) = ]-\infty,(E_1+\varepsilon_0)h]$ for $\varepsilon_0 > 0$ small enough. Another question is to find a complete expansion for $\lambda(h)$ in powers of $h$ (or in power of $h^{1/2}$ in the general case see $[SI]_2$, $[HE\text{-}SJ]_1$). The last question is to compare the eigenfunction and the approximate eigenfunction. We will answer to all these questions in the next sections.

#

## §2.3 - The B.K.W. construction.

(2.1.9) suggests strongly that we must look for a formal eigenfunction of the following type : $h^{-n/4}\, a(x,h)\, .\, e^{-\varphi(x)/h}$ which is usually called a B.K.W. solution. We shall indeed prove the following theorem :

Theorem 2.3.1. (See $[HE\text{-}SJ]_1$ for a more general result).

Under hypothesis (2.0), we can find a $C^\infty$ positive function $\varphi(x)$, a formal series :

$E(h) \sim \sum_{j \geq 1} E_j\, h^j$ $\qquad\qquad$ ($E_0 = \min V = 0$)

and a formal symbol defined in a neighborhood of 0

$$a(x,h) \sim \sum_{j=0}^{\infty} a_j(x)\, h^j$$

s.t $E_1$ is the first eigenvalue of the associate harmonic oscillator and :

$$(P(h)-E(h))\,(a(x,h)\, e^{-\varphi(x)/h}) = O(h^\infty)\, .\, e^{-\varphi(x)/h} \tag{2.3.1}$$

in a neighborhood of 0,

and :

$$a(0,h) = (2\pi)^{-n/4} \tag{2.3.2}$$

Remark 2.3.2'.

Let us recall that it is always possible to associate to a formal symbol some realization by a Borel procedure.

//

Remark 2.3.2".

The normalization we take here is not the $L^2$ normalization we use in §2. But it is easy to return to the other normalization by dividing by the formal expansion of
$\| a(x,h)\, e^{-\varphi(x)/h} \|_{L^2(B(0,\varepsilon))}$ $(\sim 1 + O(h))$
which is given by the stationnary phase theorem.

//

Remark 2.3.3.

If V is analytic in the neighborhood of 0, it can be proved (cf [HE-SJ]$_1$) that: $|E_j| \leq C^{j+1}\, j\,!$ and it is possible to find a natural procedure of summation for $(\Sigma\, E_j\, h^j)$ and $(\Sigma\, a_j(x) h^j)$ s.t

$$(P(h)-E(h))\,(a(x,h)\, e^{-\varphi(x)/h}) = \mathcal{O}(e^{-\varepsilon_0/h}) \,.\, e^{-\varphi(x)/h}$$

for some $\varepsilon_0 > 0$.

//

Remark 2.3.4.

Another interesting case is to construct B.K.W. solutions in the case where V vanishes exactly to order 2 on a smooth submanifold. This problem is studied in [HE-SJ]$_5$ and [HE-SJ]$_6$ .

//

As in the proof of proposition (2.2.1), we deduce from Theorem 2.3.1. the :

Corollary 2.3.5.

Under hypotheses (2.2.1) and (2.2.12), P(h) admits for each N at least one eigenvalue in the ball $B(\sum_{j=0}^{N} E_j\, h^j\,,\, C_N\, h^{N+1})$ for sufficiently large N and $h \in\, ]0,h_N]$.

This is also true for any Dirichlet realization of P(h) in a bounded open set containing 0.

Proof of Theorem 2.3.1.

The proof is based on the two following propositions which will be proved in section 2.4.

Proposition 2.3.6.

Under hypothesis (2.2.1), there exists a unique $C^\infty$ positive function $\varphi$ defined in a neighborhood of 0 such that

(2.3.3) $\qquad |\nabla\varphi|^2 = (V - E_0)$ $\qquad$ in this neighborhood

(2.3.4) $\varphi(x) - \varphi_0(x) = O(|x|^3)$

where $\varphi_0(x) = \sum_{j=1}^{n} \sqrt{\mu_j}\, \frac{x_j^2}{2}$

Proposition 2.3.7.

Let X be a $C^\infty$ real vector field defined in a neighborhood of 0. Suppose that the linear part of the vector fields is given by :

$$X_0 = \sum_{i=1}^{n} \nu_i\, x_i\, \partial_{x_i} \qquad \text{with } \nu_i > 0$$

Let b a $C^\infty$ function s.t $b(0) = 0$. Then for each $C^\infty$ function g s.t $g(0) = 0$ and for each $\gamma \in \mathbb{R}$, there exists a unique $C^\infty$ function f defined in a neighborhood U of 0 s.t

$$\begin{cases} (X + b)\, f = g \\ f(0) = \gamma \end{cases} \quad \text{in } U$$

To prove Theorem (2.3.1), we compute formally the left side of (2.3.1) and expand in powers of h the coefficients of $e^{-\varphi(x)/h}$.

The coefficient of $h^0$ is 0 if we have simply :

(2.3.5) $(-|\nabla\varphi|^2 + (V - E_0))\, a_0 = 0$

The equation: $|\nabla\varphi|^2 = (V - E_0)$ is called the eiconal equation.

If we take the $\varphi$ defined by proposition 2.3.6, (2.3.5) is satisfied.

Suppose now that (2.3.5) is satisfied and let us look at the coefficient of h which is 0 if :

(2.3.6) $2\, \nabla\varphi \,.\, \nabla a_0 + (\Delta\varphi - E_1)\, a_0 = 0$

and we chose the initial condition :

(2.3.7) $a_0(0) = (2\pi)^{-n/4}$

(2.3.6) is called the first transport equation.

Then we can apply proposition (2.3.7) with $f = a_0$, $\gamma = (2\pi)^{-n/4}$, $g = 0$ and $b = \Delta\varphi - E_1$, if $b(0) = 0$.

Using (2.3.4), this condition defines $E_1$ :

(2.3.8) $E_1 = \Delta\varphi(0) = \sum_{j=1}^{n} \sqrt{\mu_j}$

Let us look now at the cancellation of the coefficient of $h^2$ :

(2.3.9) $\quad 2\,\nabla\varphi\,.\,\nabla a_1 + (\Delta\varphi - E_1)\,a_1 = -\,\Delta a_0 + E_2\,a_0$

and we choose the initial condition

(2.3.10) $\quad a_1(0) = 0$

Then we can apply proposition (2.3.7) with $f = a_1$, $\gamma = 0$, $b = \Delta\varphi - E_1$ and $g = -\,\Delta a_0 + E_2\,a_0$, if $g(0) = 0$, that is :

$$-\,\Delta a_0(0) + E_2 \cdot a_0(0) = 0$$

This last equation determines $E_2$ because $a_0(0) \neq 0$ :

(2.3.11) $\quad E_2 = \dfrac{\Delta\, a_0(0)}{a_0(0)}$

The other coefficients can be treated on the same way and this finishes the proof of Theorem 2.3.1.

Remark 2.3.8.

One can also construct B.K.W. solutions attached to other levels. The proof we present here is not very well adapted to the general situation.

The best in the general case is to use our initial proof (cf [HE-SJ]$_1$) using the F.B.I. transform.

#

Remark 2.3.9.

If you are just interested to get the corollary (2.3.5), it is sufficient to work with formal series in power of x. We only need to solve :

(2.3.12) $\quad |\nabla\varphi|^2 = (V - E_0) + O(|x|^\infty)$

and the different transport equations modulo $O(|x|^\infty)$.

We need only the formal version of proposition (2.3.7) modulo $O(|x|^\infty)$.

As an exercise, the reader can verify that the $E_j$ depend only of the Taylor expansion of V at 0.

This is perhaps more transparent in the proof of Simon [SI]$_2$ .

#

Remark 2.3.10.

B.K.W. construction for from the minimum of V at a level E are of the type $a(x,h)\, e^{\pm i\,\varphi/h}$ in $\{V \leqslant E\}$ with $\varphi$ real (See Fedoryuk - Maslov [FE-MA]).

B.K.W. solutions for the Dirac operator have been constructed by X.P. Wang [WA] and in the Maslov context by Leray [LE].

§2.4 - Proof of propositions 2.3.6 and 2.3.7.

These two propositions are more or less classical in the context of the theory of the stable manifolds ([AB-MA], section 7.2).

Proof of proposition 2.3.7. (See [AB-MA], Theorem 2.1.25 for related results).

By the Sternberg's linearization theorem [BR], we can choose new coordinates y, s.t

$$y = x + 0(|x|^2)$$

and

$$X = \sum_{i=1}^{n} \nu_i \, y_i \, \partial_{y_i}$$

So we have to solve in the new coordinates

$$(2.4.1) \qquad \begin{cases} (\sum_j \nu_j \, y_j \, \partial_{y_j} + b(y))\, f = g \\ \qquad\qquad f(0) = \gamma \end{cases}$$

One reduces easily to the case $\gamma = 0$ and it is easy to solve in the category of the formal series $f \sim \sum_{|\alpha| \geq 1} a_\alpha \, y^\alpha$ under the condition (which is satisfied because $\nu_j > 0$) : $\sum_j \nu_j \alpha_j \neq 0$ for $\alpha \neq 0$.

So we are reduced to solve (2.4.1) in the space of the $C^\infty$ flat functions at 0.

Let $(t,y) \to \phi_t(y)$ the flow associated to X. It is clear that we have :

$$(2.4.2) \qquad \phi_t(y) = (e^{\nu_j t} \, y_j)_{j=1,\dots,n}$$

In particular, we get the estimate :

$$(2.4.3) \qquad |\phi_t(y)| \leq C \, e^{-\min_j \nu_j |t|} \, |y| \qquad \text{for } t \in ]-\infty, 0]$$

Then the solution of (2.4.1) (with $\gamma = 0$ and g flat) is given by :

$$(2.4.4) \qquad f(y) = \int_{-\infty}^{0} g(\phi_t(y)) \exp(- \int_t^0 b(\phi_s(y))\, ds)\, dt$$

Let us verify that the right hand side is well defined ;

If $|g(y)| \leq C_N \, |y|^N$, we have :

$$|g(\phi_t(y))| \leq C_N \, . \, C^N \, e^{-N \min_j \nu_j |t|} \, |y|^N$$

$$|- \int_t^0 b(\phi_s(y))\, ds| \leq C_0 \, |t| \, . \, |y|$$

So we get :

$$|g(\phi_t(y))\exp(-\int_t^0 b(\phi_s(y))ds)| \leq C_N . C^N |y|^N \times e^{[C_0|y|-N \min_j \nu_j]|t|}$$

If we assume that $|y| \leq \varepsilon_0$ and if we choose $N > \dfrac{C_0\,\varepsilon_0}{\min_j \nu_j} + 1$ we get that :

$$|f(y)| \leq \tilde{C}_N\, |y|^N$$

The same argument works for the derivatives of f.

Returning to the initial coordinates we get the proposition 2.3.7.

Proof of proposition 2.3.6. (Sketch)

The idea of the proof is to determine $\varphi(x)$ as the generating function of a Lagrangian manifold $\Lambda^+ = \{ \{x,\varphi'(x)\}, x \in W\}$ where W is a neighborhood of 0 lying in $q^{-1}(0)$ where $q(x,\xi) = -p(x,i\xi)\ (= |\xi|^2_x-(V-E_0))$.

In the neighborhood of (0,0) in $T^*\ \mathbb{R}^n$ (we take local coordinates (cf 2.2.1 and 2.2.2)), we have :

$$(2.4.5)\qquad q(x,\xi) = \xi^2 - \sum_j \mu_j x_j^2 + \mathcal{O}(|x,\xi|^3)$$

and

$$(2.4.6)\qquad H_q = 2\Big(\sum_j \xi_j \frac{\partial}{\partial x_j} + \sum_j \mu_j x_j \frac{\partial}{\partial \xi_j}\Big) + \mathcal{O}(|x,\xi|^2)$$

whose linear part is :

$$(2.4.7)\qquad Y_0 = 2\Big(\sum_j \xi_j \frac{\partial}{\partial x_j} + \sum_j \mu_j x_j \frac{\partial}{\partial \xi_j}\Big)$$

The associated fundamental matrix is :

$$(2.4.8)\qquad F = 2\left(\begin{array}{c|c} 0 & D \\ \hline \begin{matrix}\mu_1 & 0 \,..\, 0 \\ & \ddots \\ 0 & \mu_n\end{matrix} & 0 \end{array}\right)$$

whose eigenvalues are given by $\pm 2\sqrt{\mu_j}$ (j=1,...,n).

Let us denote by $\Lambda_0^\pm$ the positive (resp. negative) eigenspace.

It is clear that $\Lambda_0^+$ (resp. $\Lambda_0^-$) is the set of the $(x,\xi)$ s.t $\exp(-t\,Y_0)(x,\xi) \to 0$ when $t \to +\infty$ (resp. $t \to -\infty$).

Moreover, we observe that $\Lambda_0^\pm$ are Lagrangean subspaces of $T^*\ \mathbb{R}^n$, given by $\xi_j = \pm\sqrt{\mu_j}\, x_j$.

Let us recall that L is a Lagrangean Subspace of $T^*\ \mathbb{R}^n$ if it is a linear subspace of dimension n s.t $\omega(u,v) = 0$, $\forall u \in L$, $\forall v \in L$, where $\omega$ is the bilinear form on $T^*\ \mathbb{R}^n$

whose matrix in the coordinates $(x,\xi)$ is given by $\begin{pmatrix} 0 & I \\ -I & 0 \end{pmatrix}$.

Then we can apply the stable Manifold theorem for Hyperbolic Sets of Smale (See [AB-MA] th. 7.2.9) which gives the existence of two $C^\infty$ manifolds of dimension n $\Lambda^+$ and $\Lambda^-$ tangent to $\Lambda_0^+$ and $\Lambda_0^-$ at (0,0) and which can be caracterized in some adequate neighborhood of (0,0) as the set of the points $(x,\xi)$ s.t $\phi_t(x,\xi) \to (0,0)$ when t tends to $-\infty$ (resp. $+\infty$) where $\phi_t$ is the flow associated to $H_q$ .
Because $q(\phi_t(x,\xi)) = q(x,\xi)$, we get immediatly that $\Lambda^+$ and $\Lambda^-$ are contained in $q^{-1}(0)$.
Let us verify now that $\Lambda^+$ and $\Lambda^-$ are Lagrangian manifolds in $T^* \mathbb{R}^n$ (we mean by this that the tangent spaces at each point $(x,\xi)$ of $\Lambda^\pm$ are Lagrangian linear subspaces of $T_{(x,\xi)}(T^* \mathbb{R}^n)$).
If $(x,\xi) \in \Lambda^+$ and $(u,v) \in T_{x,\xi}\Lambda^+$ , we have :

$$\omega_{x,\xi}(u,v) = \omega_{\phi_t(x,\xi)}((D\,\phi_t)u,(D\,\phi_t)v)$$

(because $\phi^*_t\,\omega = \omega$ which is true because $\phi_t$ is a hamiltonian flow (prop. 3.3.4 in [AB-MA]).
Taking the limit when $t \to -\infty$, we get :

$$\omega_{x,\xi}(u,v) = 0$$

The proof of prop. 2.3.7 is now almost finished. The projection $(x,\xi) \to x$ defines a diffeomorphism of $W \cap \Lambda^+$ onto a sufficiently small neighborhood $\mathcal{V}$ of 0 in $\mathbb{R}^n$ (this is a consequence of the tangency of $\Lambda^+$ and $\Lambda_0^+$ at (0,0)).
Then we can parametrize $\Lambda^+$ as the set of the points $(x,\Psi_1(x),\ldots,\Psi_n(x))$ with $\Psi_i \in C^\infty(\mathcal{V})$.

Because $\Lambda^+$ is Lagrangean, we get : $\dfrac{\partial \Psi_i}{\partial x_j} = \dfrac{\partial \Psi_j}{\partial x_i}$ and there exists a $C^\infty$ function $\varphi$ on $\mathcal{V}$ s.t

$$\begin{cases} \dfrac{\partial \varphi(x)}{\partial x_i} = \Psi_i(x) \\ \varphi(0) = 0 \end{cases}$$

Because $T_{(0,0)}\Lambda^+ = \Lambda_0^+$ , we get also that :

$$\varphi(x) = \varphi_0(x) + \mathcal{O}(|x|^3).$$

The equation (2.3.3) is just the traduction of $\Lambda^+ \subset q^{-1}(0)$.

#

<u>Remark 2.4.1.</u> (cf Remark 2.3.9)
Let us sketch how to solve formally (2.3.12). We are looking for

$$\varphi(x) \sim \varphi_0(x) + \sum_{j \geq 1} \varphi_j(x)$$

where $\varphi_j(x)$ is an homogeneous polynomial of $d°(j+2)$.
We suppose to simplify that the metric is flat.
We write $V - E_0 = V_0(s) + \sum_{j \geq 1} V_j(s)$

where $V_j(x)$ is an homogeneous polynomial of $d°(j+2)$.
Then we get from (2.3.12) the following equations :

$$(2.4.9)_j \qquad \nabla\varphi_0 \cdot \nabla\varphi_j + \sum_{\substack{l \geq 1 \\ l+k=j}} \nabla\varphi_l \cdot \nabla\varphi_k = V_j$$

which can be solved easily by induction on j in the space $\mathcal{P}^{j+2}$ of the polynomials of degree (j+2) because $\Psi \to \nabla\varphi_0 \cdot \nabla\Psi$ is a diagonal operator on $\mathcal{P}^{j+2}$ :

$$(\nabla\varphi_0 \, \nabla.(x^\alpha) = (\Sigma \, \nu_j \, \alpha_j)(x^\alpha))$$

## §3 THE DECAY OF THE EIGENFUNCTIONS.

One of the basic tools to localize the eigenvalues will be a very precise control of the $L^2$ norms of the family of eigenfunctions $u_h$ of $P(h)$ in a fixed domain when h tends to zero. This is connected, from a technical point of view, with the control of the decay when $|x| \to \infty$ of eigenfunctions for which S. Agmon introduces his special distance. The adaptation to the semi-classical context $(h \to 0)$ has been realized by B. Simon $[SI]_3$ and in $[HE\text{-}SJ]_1$.

### §3.1 - Energy estimates.

Everything is based on the following theorem :

**Theorem 3.1.1.** ($[HE\text{-}SJ]_1$ , see also [AG], $[SI]_3$)

Let M a $C^\infty$ riemannian complete manifold and let $\Omega$ be a bounded open domain with $C^2$ boundary. Let $V \in C^0(\overline{\Omega}, \mathbb{R})$ and $\phi$ a real valued Lipschitzian function on $\overline{\Omega}$ ($\nabla\phi$ is in this case well defined in $L^\infty(\Omega)$, where $\nabla$ is the gradient operator).

Then, for any $u \in C^2(\overline{\Omega}, \mathbb{R})$ with $u_{\restriction \partial\Omega} = 0$, we have :

$$(3.1.1) \qquad h^2 \int_\Omega |\nabla(e^{\phi/h} u)|^2_x \, dx + \int_\Omega (V - |\nabla \varphi(x)|^2_x) \, e^{2\phi/h} u^2 \, dx = \int_\Omega e^{2\phi/h} \, Pu.u \, dx$$

where $P = -h^2 \Delta + V$ , $h > 0$, dx is the Riemannian volume, $|\ |_x$ is the norm in $T_x M$.

**Proof**

In the case where $\phi$ is a $C^2(\overline{\Omega})$-function, this is an immediate consequence of the Green-formula :

$$(3.1.2) \qquad \int_\Omega |\nabla v|^2 \, dx = -\int_\Omega \Delta v.v \, dx$$

for $v \in C^2(\overline{\Omega})$ s.t $v_{\restriction\partial\Omega} = 0$

To treat the general case, we just write $\phi$ as a limit as $\varepsilon \to 0$ of $\phi_\varepsilon = \chi_\varepsilon * \phi$ (we write the argument in the case of $\mathbb{R}^n$) where $\chi_\varepsilon(x) = \chi(\frac{x}{\varepsilon})\, \varepsilon^{-n}$ is a standard mollifier and we remark that $\nabla\phi$ is almost everywhere the limit of $\nabla\phi_\varepsilon = \nabla\chi_\varepsilon * \phi$ (in the Riemannian case write $\phi = \Sigma\, \phi_j$ with supp $\phi_j$ in an open chart $O_j$ and define $\phi_j^\varepsilon$ in the system of cordinates ! ).

### §3.2. The Agmon distance.

Suppose that we are interested in the study of eigenvalues of P(h) (or of a Dirichlet realization of P(h) in some open set $\Omega$) living in the neighborhood of E (with $\lim_{|x| \to \infty} V > E \geq \min V$ in the case $M = \mathbb{R}^n$, $E \geq \min V$ in the case of M compact or in the case of $\overline{\Omega}$ compact).

The Agmon metric is defined as : $(V-E)_+ \, dx^2$ where $dx^2$ is the Riemannian metric in

the manifold M. The Agmon metric is nothing else than the Jacobi metric introduced in classical mechanics in the set $V \leq E$, but we have replaced (E-V) by (V-E) because we want to work in the classically forbidden domain : $V \geq E$.
We refer to section (3.7) in [AB-MA] (particularly Def 3.7.6. and Theorem (3.7.7)) for the study of the Jacobi metric, and we get most of the properties of the Agmon metric from the properties of the Jacobi metric by replacing (E-V) by (V-E). The Agmon metric is degenerate in the well : $V \leq E$. Associated to the Agmon metric, we define a natural distance $d(x,y)$ (which is also degenerate) which is the infimum of the length of piecewise $C^1$ paths connecting x and y.
We have the following classical properties :

(3.2.1) $$\begin{cases} \forall\, x,x',y \in M \\ |d(x',y) - d(x,y)| \leq d(x',x) \end{cases}$$

(3.2.2) $$|\nabla_x d(x,y)|^2 \leq (V-E)_+ (x) \qquad \text{for almost every } x$$

(3.2.2) is also verified for any distance : $d(x,U) = \inf\limits_{y \in U} d(x,y)$.

---

Let us now consider the particular case of a non degenerate well where :

(3.2.3) $$V(0) = V'(0) = 0 \quad , (\text{Hess } V)(0) > 0 \quad , \quad E = 0$$

We have the following lemma :

Lemma 3.2.1.

Under hypothesis (3.2.3), then

$$d(x,0) = \varphi(x) \qquad \text{in a neighborhood of } 0$$

where $\varphi(x)$ was introduced in proposition 2.3.6.

Proof (we write the proof in the case of $R^n$ to simplify)

(a) $d(x,0) \geq \varphi(x)$

Let $\gamma : [a,b] \ni t \to x(t)$ a piecewise $C^1$ path joining 0 and x.

Then there exists $a < c \leq b$ s. t

$$\begin{cases} \varphi(x(t)) \leq \varphi(x) & \text{for } t \in [a,c] \\ \varphi(x(c)) = \varphi(x) \end{cases}$$

Then $\varphi(x) = \int_a^c \frac{d}{dt}\varphi(x(t))\,dt = \int_a^c \varphi'(x(t)).\dot{x}(t)\,dt$

$$\leqslant \int_a^c |\varphi'(x(t))|\;|\dot{x}(t)|\,dt \leqslant \int_a^c \sqrt{V(x(H))}\;|\dot{x}(t)|\,dt$$

$$\leqslant \int_a^b \sqrt{V(x(t))}\;|\dot{x}(t)|\,dt = \text{length}(\gamma)$$

$$\varphi(x) \leqslant \text{length } \gamma$$

Taking the infimum over $\gamma$, we get the result.

(b) $\varphi(x) \geqslant d(x,0)$

Let x in the neighborhood of 0 $\Omega$ s.t $\Lambda_+ = \{(x,\varphi'(x)),\ x \in \Omega\}$ (cf proof of proposition 2.3.6, §2.4).

By the construction of $\Lambda_+$, there exists a bicaracteristic curve of $H_q$ $(x(t),\xi(t))$ parametrized by $]-\infty,0]$ s.t.

$$(x(0),\xi(0)) = (x,\varphi'(x))$$

$$(x(-\infty),\xi(-\infty)) = (0,0).$$

Let us remark now that :

$$\frac{d}{dt}\varphi(x(t)) - \varphi'_x(x(t)).\dot{x}(t) - 2\,\xi(t).\xi(t)$$

$$= 2\|\xi(t)\|\;\|\xi(t)\| = \sqrt{V(x(t))}\;\|\dot{x}(t)\|$$

Then it is clear that :

$$\varphi(x) = \int_{-\infty}^{0} \sqrt{V(x(t))}\;|\dot{x}(t)|\,dt$$

Then $\varphi(x)$ appears us the length of the semi-infinite path $]-\infty,0] \ni t \to x(t)$.

Let $\gamma_\varepsilon$ the path defined by :

$$\gamma_\varepsilon(t) = \frac{t}{\varepsilon}\,x\left(\frac{-1}{\varepsilon}\right) \qquad \text{for } 0 < t < \varepsilon$$

$$\gamma_\varepsilon(t) = x\left(\frac{-1}{\varepsilon} + (t-\varepsilon)\right) \qquad \text{for } \varepsilon < t \leqslant +\frac{1}{\varepsilon} + \varepsilon$$

then it is clear that leigth $(\gamma_\varepsilon) \xrightarrow[\varepsilon \to 0]{}$ length $\gamma$ so we get $d(x,0) \leqslant \varphi(x) + o(\varepsilon)$, $\forall \varepsilon > 0$.

§3.3 - Decay of eigenfunctions for the one-well Dirichlet Problem.

We shall see in §4 that an important step is the full understanding of the problem with one well. Let us assume in this section that :

$$U = \{x \in M,\ V(x) \leq E\}$$

is compact (take $E < \varliminf_{|x| \to \infty} V$ if $M = \mathbb{R}^n$) and has diameter 0 for the Agmon distance.

Let us consider some bounded regular open set $\Omega$ containing U. Then we can define the Dirichlet realization of the Schrödinger operator in $\Omega$: $P_\Omega(h)$ whose domain is $H^2(\Omega) \cap H^1_0(\Omega)$. Let $\mathfrak{J}$ some subset of $]0,h_0]$ admitting 0 as a point of accumulation and suppose that we have a family of eigenfunctions $u_h$ s.t

$$(3.3.1)\quad \begin{cases} P_\Omega(h).u_h(x) = (E+\lambda(h))\, u_h(x) \\ \|u_h\| = 1 \\ \max(\lambda(h),0) \underset{\substack{h \to 0 \\ h \in \mathfrak{J}}}{\longrightarrow} 0 \end{cases}$$

Then we have (with $d(x,U) = d_{V-E}(x,U)$, the Agmon distance associated with $(V-E)$) :

Proposition 3.3.1.

Under Hypothesis (3.3.1), we have :

For each $\varepsilon > 0$, there exists $C_\varepsilon > 0$ s.t for $h \in ]0,h_0]$

$$(3.3.2)\quad \|\nabla(e^{d(x,U)/h} u_h)\|_{L^2(\Omega)} + \|u_h\, e^{d(x,U)/h}\|_{L^2(\Omega)} \leq C_\varepsilon\, e^{\varepsilon/h}$$

Corollary 3.3.2.

The $L^2$ norm of $u_h$ is concentrated in the well, that means that, for each open neighborhood $\mathcal{V}$ of U in $\Omega$, we have :

$$(3.3.3)\quad \|u_h\|_{L^2(\mathcal{V})} = 1 + \mathcal{O}(e^{-\varepsilon/h}) \qquad \text{with } \varepsilon > 0$$

Proof

Let us choose some $\varepsilon > 0$. We shall use the identity (3.1.1) with :

$$\begin{cases} V \text{ replaced by } V-(E+\lambda(h)) \\ \phi(x) = (1-\delta)\, d(x,U) \quad \text{(with } \delta \text{ to be chosen sufficiently small (depending on } \varepsilon)) \\ u = u_h \\ P = -h^2 \Delta + V-(E+\lambda(h)) \end{cases}$$

Indeed this identity is also valid for $u \in D(P_\Omega)$.

Let $\qquad \Omega_\delta^+ = \{x \in \Omega,\ V(x) \geq \delta + E\}$

$$\Omega_\delta^- = \{x \in \Omega,\ V(x) < \delta + E\}$$

Then we deduce from (3.1.1) the inequality : $\forall u \in D(P_\Omega)$

$$h^2 \int_\Omega |\nabla(e^{\phi/h} u_h)|^2\, dx + \int_{\Omega_\delta^+} (V-E-\lambda(h)-|\nabla\phi|^2)\, e^{2\phi/h} u_h^2\, dx$$

$$\leq \sup_{x \in \Omega_\delta^-} |V-E-\lambda(h)-|\nabla\varphi|^2| \left(\int_{\Omega_\delta^-} e^{2\phi/h} u_h^2\, dx\right)$$

and then, for some constant C independent of $h \in ]0,h_0]$, $\delta \in [0,1]$ :

(3.3.4) $$h^2 \int_\Omega |\nabla(e^{\phi/h} u_h)|^2\, dx + \int_{\Omega_\delta^+} (V-E-\lambda(h)-|\nabla\phi|^2)\, e^{2\phi/h} u_h^2\, dx \leq C\left(\int_{\Omega_\delta^-} e^{2\phi/h} u_h^2\, dx\right)$$

Let us observe now that on $\Omega_\delta^+$ we have (with $\phi = (1-\delta)d(x,U)$) by (3.2.2) :

$$V-E-\lambda(h)-|\nabla\phi|^2 \geq (1-(1-\delta)^2)(V-E)-(\lambda(h))_+$$

$$\geq \delta\,[2-\delta]-(\lambda(h))_+ \qquad \text{a.e}$$

For $h \in ]0,h(\delta)]$ where $h(\delta)$ is determined by :

$$\lambda_+(]0,h(\delta)]) < \delta^2$$

we get :

$$h^2 \int_\Omega |\nabla(e^{\phi/h} u_h)|^2 dx + \delta^2 \int_{\Omega_\delta^+} e^{2\phi/h} u_h^2\, dx \leq C \int_{\Omega_\delta^-} e^{2\phi/h} u_h^2\, dx$$

and then

$$h^2 \int_\Omega |\nabla(e^{\phi/h} u_h)|^2 dx + \delta^2 \int_\Omega e^{2\phi/h} u_h^2\, dx$$

$$\leq (C+1) \int_{\Omega_\delta^-} e^{2\phi/h} u_h^2\, dx$$

$$\leq (C+1)\, e^{\frac{2}{h}\left(\sup_{x \in \Omega_\delta^-} \phi\right)} \qquad \text{(because } \|u_h\|^2 = 1)$$

$$\leq (C+1)\, e^{a(\delta)/h} \qquad \text{with } a(\delta) \to 0 \text{ for } \delta \to 0$$

We have now to get the same type of estimates for $\phi$ replaced by $d(x,U)$. Let us observe first that :

$$h^2 \int_\Omega |\nabla(e^{d(x,U)/h} u_h)|^2 dx = h^2 \int_\Omega |\nabla(e^{\delta . \frac{d(x,U)}{h}} . e^{(1-\delta)\frac{d(x,U)}{h}} u_h)|^2 dx$$

$$\leqslant h^2 e^{\frac{2\delta M}{h}} \int_\Omega |\nabla e^{\phi(x)/h} u_h|^2 dx$$

$$+ \delta^2 (\sup_{x \in \Omega} |V-E|) (\int_\Omega e^{2\phi/h} u^2_h dx).e^{\frac{2\delta M}{h}}$$

It is then clear that we get :

$$h^2 \int_\Omega |\nabla(e^{d(x,U)/h} u_h)|^2 + \delta^2 \int_\Omega e^{2d(x,U)/h} u^2_h dx \leqslant (C+1)[1+ \sup_{x \in \Omega} |V-E|] e^{\frac{a(\delta)+2\delta M}{h}}$$

Then we can choose $\delta$ s.t $a(\delta) + 2\delta M < \varepsilon/2$ and we get (3.3.2).

#

Remark 3.3.3.

Proposition (3.3.1) gives you a good control of $e^{d(x,U)/h} u_h$ in the norm $H^1$. It is sometimes useful to have pointwise estimates. For given $\varepsilon$, we can regularize $d(x,U)$ to get a $C^\infty$ function $\phi_\varepsilon(x)$ on $\overline{\Omega}$ s.t :

(3.3.5) $$\sup_{x \in \overline{\Omega}} |d(x,U) - \phi_\varepsilon(x)| < \varepsilon$$

Then we get from (3.3.2) the estimate :

$$\| e^{\phi_\varepsilon(x)/h} u_h \|_{H^1} \leqslant \tilde{C}_\varepsilon e^{3\varepsilon/h}$$

Let us now compute :

$$(-h^2 \Delta + (V-E)-\lambda(h))(e^{\phi_\varepsilon(x)/h} u_h) = -h\, 2(\nabla \phi_\varepsilon)(\nabla u_h)\, e^{\phi_\varepsilon(x)/h} - h^2(\Delta(e^{\phi_\varepsilon(x)/h})) u_h$$

So we get the system

$$\begin{cases} -\Delta(e^{\phi_\varepsilon(x)/h} u_h) = \frac{1}{h^2} a_\varepsilon(x,h)\, e^{\phi_\varepsilon(x)/h} u_h + \frac{1}{h^2} \vec{b}_\varepsilon(x,h).\nabla(e^{\phi_\varepsilon(x)/h} u_h) \\ e^{\phi_\varepsilon(x)/h} u_{h/\partial\Omega} = 0 \end{cases}$$

The classical regularity theorem for the Dirichlet realization of $-\Delta$ in $\Omega$ then gives :

$$\| e^{\phi_\varepsilon(x)/h} u_h \|_{H^{k+2}} \leqslant \frac{C}{h^2} \| e^{\phi_\varepsilon(x)/h} u_h \|_{H^{k+1}}$$

and finally :

$$\| e^{\phi_\varepsilon(x)/h} u_h \|_{H^{k+2}(\Omega)} \leqslant \frac{C_\varepsilon}{h^{2k}} \cdot e^{3\varepsilon/h} \leqslant C_{\varepsilon,k}\, e^{4\varepsilon/h}$$

Using the Sobolev embedding theorem, we get the following estimate :
$\forall \varepsilon > 0$, $\forall \alpha$ $\exists C_{\alpha,\varepsilon}$ s.t $\forall h \in ]0,h_0]$, $\forall x \in \Omega$ :

$$|D_x^\alpha u_h(x)| \leqslant C_{\alpha,\varepsilon}\, e^{4\varepsilon/h} \,.\, e^{-\phi_\varepsilon(x)/h}$$

and finally using (3.3.5) we get the :

Proposition 3.3.4. (we write the case $M = \mathbb{R}^n$, to simplify).
Under hypothesis (3.3.1), for each $\varepsilon > 0$, each $\alpha \in \mathbb{N}^n$, $\exists C_{\varepsilon,\gamma}$ s.t, for each $h \in ]0,h_0]$, each $x \in \Omega$, we have :

(3.3.6) $$|D_x^\alpha u_h(x)| \leqslant C_{\varepsilon,\alpha}\, e^{\varepsilon/h} \,.\, e^{-d(x,U)/h}$$

The propositions (3.3.1) and (3.3.4) can be improved in the case when the well is non-degenerate and when $E = \min V$. So we introduce the following hypothesis :

(3.3.7) $V(0) = E = 0$, $V'(0) = 0$, $V''(0) \geqslant 0$.

(3.3.8) $\lambda(h) \in [0,C_0\, h]$

In this context we get the :

Proposition 3.3.5. (See Prop. 5.5 in [HE-SJ]$_1$)
Under hypotheses (3.3.1), (3.3.7), (3.3.8), we get the existence for each $\alpha \in N^n$ of $C_\alpha > 0$ and $N_\alpha \in \mathbb{N}$ s.t

(3.3.9) $$|D_x^\alpha u_h(x)| \leqslant C_\alpha\, h^{-N_\alpha}\, e^{-d(x,U)/h} \quad \text{in } \Omega_0$$

where $\Omega_0$ is a neighborhood of 0.

The proof is more or less the same as before with the following modifications. Set $d(x) = d(x,0)$ ; we take :

$$\phi(x) = d(x) - C\, h\, \mathrm{Log}(d(x)/h) \quad \text{for } d(x) \geqslant C\, h$$

$$= d(x) - C\, h\, \mathrm{Log}\, C \qquad \text{for } d(x) \leqslant C\, h$$

$$\Omega_+ = \{x, d(x) \geq C\,h\}$$

$$\Omega_- = \{x, d(x) \leq C\,h\}$$

where C is a positive constant depending of $C_0$ (in 3.3.8) and of $C_1$ where $C_1$ is the constant s.t :

$$\frac{1}{C_1} \leq \frac{V(x)}{d(x)} \leq C_1$$

(See prop. 2.3.6. for the property of d(x) in the neighborhood of 0 and lemma (3.2.1)).

## §3.4 - Localiztion of the Spectrum for the Dirichlet Problem at the bottom.

We have seen in section (2.3) (Corollary 2.3.5) that, under hypothesis (3.2.3), there exists at least one eigenvalue for the Dirichlet realization $P_\Omega(h)$ near the eigenvalue of the localized harmonic oscillator at 0 : $E_0 + E_1\,h$. We shall prove in this section that, in fact, for $\varepsilon > 0$ small enough and for $h \in ]0,h_0(\varepsilon)]$ there is only one eigenvalue of $P_\Omega(h)$ in $[E_0, E_0+E_1 h+\varepsilon h]$ (which appears to be of multiplicity one ! ). More precisely we shall prove the :

Theorem 3.4.1. [See $[SI]_2$, $[HE\text{-}SJ]_1$]

Let $\Omega \ni 0$ an open, sufficiently small, regular bounded set.
Under hypothesis (2.0), $P_\Omega(h)$ admits exactly one simple eigenvalue in the ball $B(E_0+E_1 h, C_1\,h^2)$ (for $C_1$ big enough and $h \in ]0,h_0(C_1)]$) and this eigenvalue admits as an asymptotic expansion in powers of h the serie E(h) determined in Theorem 2.3.1.

Proof

Essentially, we prove his result by deforming the initial Schrödinger Operator onto an harmonic oscillator for which the spectrum is well known.

Step 1

The first step is to give an inequality in the spirit of the study of the Grusin's type hypoelliptic operators.

Lemma 3.4.2.

Under hypothesis (3.2.3) and for $\Omega$ sufficiently small there exists, for every $C_0 > 0$ and every $M \in \mathbb{N}$, a constant $C_M > 0$ s.t for all $E \in [0, C_0\,h]$, $h \in [0,h_0[$, $u \in C_0^\infty(\Omega)$ :

(3.4.1) $$\|u\|_{M+2} \leq C_M \left(\left\| \frac{1}{h}(P(h)-E)u\right\|_M + \|u\|_0\right)$$

Here :

$$(3.4.2) \qquad \|u\|_M \overset{\text{définition}}{\equiv} \sum_{|\alpha+\beta|\leqslant M} h^{-(|\alpha|+|\beta|)/2} \|x^\alpha (hD_x)^\beta u\|_{L^2}$$

If we suppose now that $E \in [(E_1+\varepsilon)h,(E_1+2\varepsilon)h]$ where $\varepsilon$ is chosen $> 0$ s.t $[(E_1+\varepsilon),(E_1+2\varepsilon)]$ does'nt meet the spectrum of the localized Harmonic oscillator, then we have for $h \in [0,h_1(\varepsilon)[$ :

$$(3.4.3) \qquad \|u\|_{M+2} \leqslant \tilde{C}_M(\varepsilon) \, \| \tfrac{1}{h} (P(h)-E)u\|_M$$

Proof

To simplify the notations, we just prove the case where the manifold is $\mathbb{R}^n$. Let us start with the proof of $(3.4.1)_0$.

As we have seen in section 2, we can write :

$$P(h) = P_0(h) + (V-V_0) \qquad \text{with } (V-V_0)(x) = 0(|x|^3)$$

Starting from the inequality $(3.4.1)_0$ for $P_0(h)$ which is easy to get (first prove it for $h = 1$ by expanding $u$ in the orthonormal basis of the Hermite functions for $P_0(1)$ and then prove it in the general case by a scaling argument), we get $(3.4.1)_0$ in the general case by observing that :

$$\|(V-V_0)\, u\|_0 \leqslant C \sup_{x\in\Omega} |x|\, h \, \|u\|_2 \ , \qquad \forall u \in C_0^\infty(\Omega)$$

Then we get $(3.4.1)_0$ for $P(h)$ by taking $\Omega$ sufficiently small. The general inequality for $M > 0$ is simply proved by induction on $M$.

# End of the proof of lemma 3.4.2.

Step 2

Let us consider now $u_h$ a normalized eigenfunction of $P_\Omega(h)$ associated to an eigenvalue $\lambda(h) \in [0,C_0\, h]$. Let $\chi \in C_0^\infty(\Omega)$ be equal to 1 near 0 and define : $v_h = \chi u_h$ .
By the decay properties of $u_h$ (See corollary 3.3.2), it is clear that

$$(3.4.4) \qquad \|u_h - v_h\|_{L^2(\Omega)} = \mathcal{O}(e^{-\varepsilon_0/h}) \qquad \text{for some } \varepsilon_0 > 0$$

and that :

$$(3.4.5) \qquad P(h)\, v_h = \lambda(h)\, v_h + \mathcal{O}(\exp^{-\varepsilon_0/h})$$

By lemma (3.4.2) we get that :

(3.4.6) $\quad \| v_h \|_{M+2} \leq C_M \, , \qquad \forall M \in \mathbb{N}$

On the other hand :

$$P_0(h).v_h = \lambda(h)\, v_h + (V_0 - V)\, v_h + \mathcal{O}(e^{-\varepsilon_0/h})$$

(3.4.7) $\quad P_0(h)\, v_h = \lambda(h)\, v_h + \mathcal{O}(h^{3/2}) \qquad$ (by 3.4.6 with M = 1)

As in §2, we deduce from (3.4.4) and (3.4.7) that for some constant $\tilde{C} > 0$ and for $h \in ]0,h_0]$ :

(3.4.8) $\quad Sp(P_\Omega(h)) \cap [0, C_0\, h] \subset Sp(P_0(h)) + B(0, \tilde{C}\, h^{3/2})$

where $\Omega$ is a sufficiently small open set containing 0 (in fact, if you analyze more precisely this proof you get also (3.4.8) without any restriction on $\Omega$). (3.4.8) does'nt give the exact number of eigenvalues of $P_\Omega$ contained in $[Sp\, P_0(h) + B(0,\tilde{C}\, h^{3/2})] \cap [0, C_0\, h]$ so we must be more precise in the argument. This is the subject of :

Step 3 (the deformation argument).
Let $P_\Omega^{(t)}(h)$ be the Dirichlet realization in $\Omega$ of :

$$P^{(t)}(h) = -h^2\, \Delta + V_0 + (1-t)(V - V_0)$$

Then, following the arguments of step 1 and step 2, it is easy to see that the spectrum of $P_\Omega^{(t)}(h)$ does'nt meet the interval $[(E_1+\varepsilon)h, (E_1+2\varepsilon)h]$ and that the number of eigenvalues of $P_\Omega^{(t)}(h)$ less than $(E_1+\varepsilon)h$ is constant in the deformation from t = 0 to t = 1.
We are then reduced to count the number of eigenvalues of $P_\Omega^{0}(h)$ less than $(E_1+\varepsilon)h$.

Step 4
The last problem is to prove the :

Lemma 3.4.3
Let $0 \in \Omega \subset \Omega' \subset M$ and suppose that there is only one non degenerate well at 0 in $\Omega'$ s.t $V(0) = 0$; then, if $C_0$ avoid an eigenvalue of $P_0(1)$, there exists a bijection b of $Sp(P_\Omega(h)) \cap [0, C_0\, h]$ onto $Sp(P_{\Omega'}(h)) \cap [0, C_0\, h]$ respecting the multiplicities s.t
$b(\lambda) - \lambda = \mathcal{O}(e^{-\varepsilon_0/h})$

This lemma will be proved in §4 (Remark 4.3.3).
Appling this lemma with $\Omega = \Omega$, $\Omega' = \mathbb{R}^n$ and $P(h) = P_0(h)$ we get the proof because the spectrum of $P_0(h)$ is well known.

Remark 3.4.4.

From lemma (3.4.3), we get also that theorem 3.4.1 is true for arbitrary $\Omega$ containing a unique non degenerate well.

Remark 3.4.5.

A more direct proof of Theorem 3.4.1 is given in $[SI]_2$. But the proof we give here can be adapted to treat cases when the minima are submanifolds (see $[HE\text{-}SJ]_5$, $[HE\text{-}SJ]_6$.)

§3.5 - Decay in the case of multiple wells Dirichlet problems.

Let us just point out in this section the results we get on the decay of eigenfunctions in the case when

$$V^{-1}\ (]-\infty,E]) = \bigcup_j U_j$$

for the Dirichlet realization $P_\Omega(h)$ in an open bounded set containing $V^{-1}(]-\infty,E])$. Here the $U_j$ are compact disjoint sets of diameter 0 for the Agmon-distance. Then we have the following results.

Proposition 3.5.1.

Proposition 3.3.1 and Proposition 3.3.4 are true with :

$$d(x,U) = \inf_j d(x,U_j)$$

Proposition 3.5.2.

If M is a compact Manifold, Propositions 3.3.1 and 3.3.4 are true for the eigenfuncitons of P(h) in M.

Proposition 3.5.3.

If $M = \mathbb{R}^n$ and $E < \lim\limits_{|x|\to\infty} V$, we have under hypothesis (3.3.1) and for each $\varepsilon > 0$ the following inequality :

$$\|\nabla(e^{(1-\varepsilon)d(x,U)/h}\, u_h)\|_{L^2(\mathbb{R}^n)} + \|u_h\, e^{(1-\varepsilon)d(x,U)/h}\|_{L^2(\mathbb{R}^n)} = \mathcal{O}(e^{\varepsilon/h})$$

Moreover (3.3.6) is true on each compact of $\mathbb{R}^n$.

Sketch of the Proof

We first make the proof with $\phi_R(x) = (1-\delta)\ \chi_R(d(x,U))$ where $\chi_R(t) = t$ for $t \in [0,R]$, $\chi_R(t) = R$ for $t \geq R$ and then we take the limit $R \to \infty$.

## §4 STUDY OF INTERACTION BETWEEN THE WELLS.

### §4.1 - Preparations in functional Analysis.

Let us first discuss some more or less classical notions on distance between closed subspaces of a Hilbert space.

These notions are used in numerical Analysis but we refer to [HE-SJ]$_1$ for detailed proofs of some basic facts we recall now.

Let E, F be closed subspaces in a Hilbert space H. Let $\Pi_E$, $\Pi_F$ be the orthogonal projections on E and F respectively. We then define the non-symmetric distance $\vec{d}(E,F)$ as:

$$(4.1.1) \qquad \vec{d}(E,F) = \| \Pi_E - \Pi_F \Pi_E \| = \| \Pi_E - \Pi_E \Pi_F \|$$

Note that :

$$(4.1.2) \qquad \vec{d}(E,F) = 0 \qquad \text{if and only if } E \subseteq F$$

and that :

$$(4.1.3) \qquad \vec{d}(E,G) \leqslant \vec{d}(E,F) + \vec{d}(F,G)$$

if G is a third closed space of H.

Then we shall use the following properties :

(4.1.4) If $\vec{d}(E,F) < 1$, then $\Pi_{F/E} : E \to F$ is injective

and $\Pi_{E/F} : F \to E$ has a continuous right inverse

(4.1.5) If $\vec{d}(E,F) < 1$ and $\vec{d}(F,E) < 1$, then $\Pi_{E/F} : F \to E$

and $\Pi_{F/E}$ are bijective with continuous inverse. Moreover $\vec{d}(E,F) = \vec{d}(F,E)$.

The following proposition will be very useful to compare different spectra of selfadjoint operators.

Proposition 4.1.1. (cf [HE-SJ]$_1$)

Let A be a self adjoint operator in a Hilbert space H.

Let $I \subseteq \mathbb{R}$ be a compact interval, $\Psi_1, \dots, \Psi_N \in H$ linearly independents in D(A) and $\mu_1, \dots, \mu_N \in I$ such that

$$(4.1.6) \qquad A\,\Psi_j = \mu_j . \Psi_j + r_j \qquad \text{with } \|r_j\| \leqslant \varepsilon$$

Let $a > 0$ and assume that $Sp(A) \cap (I + B(0,2a) \setminus I) = \emptyset$.

Then if E is the space spanned by $\Psi_1, \dots, \Psi_N$ and if F is the space associated to Sp(A) $\cap$ I, we have

$$\vec{d}(E,F) \leqslant \frac{N^{1/2}\,\varepsilon}{a(\lambda_S^{min})^{1/2}} \tag{4.1.7}$$

where $\lambda_S^{min}$ is the smallest eigenvalues of $S = ((\Psi_j/\Psi_k))$.

Remark 4.1.2.

If Sp(A) $\cap$ I is discrete of finite multiplicity and if the right hand side of (4.1.7) is strictly smaller than 1, then we conclude that A has at least N eigenvalues in I.

§4.2 - To what extent can we forget the tunneling effect ?

In this subsection, we will essentially prove that, modulo an exponentially small error, the spectrum of the Schrödinger operator in some intervall I(h) is the same as the spectrum of the direct sum of Dirichlet one well problems in I(h). Let us take E = 0 and suppose that :

$$\min V \leqslant 0 < \varliminf_{|x| \to \infty} V \tag{4.2.1}$$

(the last condition, only if $M = \mathbb{R}^n$)

and write the decomposition of $V^{-1}(]-\infty,0])$ :

$$V^{-1}(]-\infty,0]) = U_1 \cup U_2 \dots \cup U_N \tag{4.2.2}$$

and suppose that the $U_j$ are disjoint, compact and s.t $\delta(U_j) = 0$.

The $U_j$ are called the wells. $\delta$ is the diameter associated to the Agmon metric $V^+ dx^2$ and we denote by d(x,y) the associate distance. Let us denote by $S_0$ the minimal distance between the different wells :

$$S_0 = \min_{j \neq k} d(U_j,U_k) \tag{4.2.3}$$

We shall associate to each well a Dirichlet problem in some open set $M_j$ containing $U_j$.

Construction of $M_j$ in the case when M is compact.

Let $B(U_j,\eta) = \{x \in M,\ d(x,U_j) \leqslant \eta\}$. Then we define :

$$M_j^{(\eta)} = M \setminus \bigcup_{k \neq j} B(U_k,\eta) \qquad \text{(with } \eta > 0\text{, small enough)}$$

In the case when $\partial M_j^{(\eta)}$ is not $C^2$, it is always possible to modify $M_j^{(\eta)}$ in $\check{M}_j^{(\eta)}$ to have a $C^2$ boundary and :

$$M_j^{(\eta)} \subset \check{M}_j^{(\eta)} \subset M_j^{(\eta/2)}$$

Construction in the case of $\mathbb{R}^n$.

We just take :

$$\tilde{M}_j^{(\eta)} = M_j^{(\eta)} \cap \overset{\circ}{B}(U_j,S) \qquad \text{for some large } S > 2S_0$$

If necessary we regularize the boundary.

In all the cases, we write now $M_j^{(\eta)}$ or more simply $M_j$ (but don't forget that everything depends on a smooth way of $\eta$).

Let $P_{M_j}(h)$ denote the Dirichlet realization of $P(h)$ in $M_j$. Let us choose now some subset $\mathcal{J} \subset ]0,h_0]$ s.t 0 is a point of accumulation for $\mathcal{J}$.

Let

(4.2.4) $\quad I(h) = [\alpha(h),\beta(h)] \quad (h \in \mathcal{J})$ be an intervall s.t :

$$\alpha(h) \to 0, \ \beta(h) \to 0.$$

and $a(h)$ some function on $\mathcal{J}$ s.t :

(4.2.5) $\quad |\log a(h)| = o(1/h)$

Then , let us write :

$$(4.2.6) \quad \begin{cases} (1) \quad \operatorname{Sp} P(h) \cap I(h) = \{\lambda_1, \ldots, \lambda_M\} \\ (2) \quad \operatorname{Sp} P_{M_j}(h) \cap I(h) = \{\mu_{j,1}, \ldots, \mu_{j,m_j}\} \end{cases}$$

and let us recall from §1 that :

(4.2.7) $\quad M + m_1 + \ldots + m_N = \mathcal{O}(h^{-n})$

Let us denote by $U_k$ (resp $\varphi_{j,k}$) orthonormalized eigenfunctions associated to $\lambda_k$ (resp. $\mu_{j,k}$).

---

Let $\theta_j$ be a $C^\infty$ function with support in $B(U_j,2\eta)$ and equal to 1 an $B(U_j, \frac{3}{2}\eta)$. Then we define :

(4.2.8) $\quad \chi_j = 1 - \sum_{k \neq j} \theta_k \qquad$ if M is compact

(4.2.8)' $\chi_j = (1 - \sum_{k \neq j} \theta_k)\, \chi_S^j$ if $M = \mathbb{R}^n$

(where $\chi_S^j$ is with support in $B(U_j,S)$ and equal to 1 in $B(U_j,S-\eta)$)
and let us introduce :

(4.2.9) $\Psi_{j,k} = \chi_j \cdot \varphi_{j,k}$

The aim of this section is to prove the following :

Theorem 4.2.1.

Let $E_j$ the space spanned by the $\Psi_{j,k}$ $(k=1,\dots,m_j)$, $E = \bigoplus_j E_j$ and let F the eigenspace of $P(h)$ attached to $I(h)$ $(h \in \mathfrak{J})$. Let us suppose that :

$$(4.2.10)\quad \begin{cases} Sp(P(h)) \cap (I(h) + B(0,2a(h)) - I(h)) = \emptyset \\ Sp\, P_{M_j}(h) \cap (I(h) + B(0,2a(h)) - I(h)) = \emptyset \qquad \text{for } h \in \mathfrak{J}\ (j=1,\dots,N) \end{cases}$$

then, for each $\sigma < S_0 - 2\eta$, we have :

(4.2.11) $\vec{d}(E,F),\ \vec{d}(F,E) = \mathcal{O}(e^{-\sigma/h})$

Moreover, for h sufficiently small, there exists a bijection b from $Sp(P) \cap I$ (where we count the eigenvalues with their multiplicity ! ) onto

$$\bigcup_{j=1}^{N} (Sp(P_{M_j}(h)) \cap I(h)$$

s.t, for every $\sigma < S_0 - 2\eta$, we have :

$$b(\lambda) - \lambda = \mathcal{O}(e^{-\sigma/h})$$

Remark 4.2.2.

In the future, we will write sometimes (4.2.11) under the following form :

$$\vec{d}(E,F) = \tilde{\mathcal{O}}(e^{-S_0/h})$$

That means that for each $\varepsilon > 0$, we can choose $\eta_0 > 0$ s.t for each $0 < \eta < \eta_0$, we get :

$$\vec{d}(E^{(\eta)},F) = \mathcal{O}_\eta(e^{-\frac{S_0}{h} + \frac{\varepsilon}{h}})$$

Similarly, we will write :

$$b(\lambda) - \lambda = \tilde{\mathcal{O}}(e^{-\frac{S_0}{h}})$$

That means that for $0 < \eta < \eta_0$, we can find $b^{(\eta)}$ s.t

$$b^{(\eta)}(\lambda) - \lambda = \mathcal{O}(e^{-\frac{S_0}{h} + \frac{\varepsilon}{h}})$$

#

Proof of theorem 4.2.1.

a) Proof of $\vec{d}(E,F) = \mathcal{O}(e^{-\sigma/h})$

We just apply proposition 4.1.1. We just observe, as a consequence of the choice (4.2.8) of $\chi_j$ and of (3.3.6) applied to $\varphi_{j,k}$, that we have :

(4.2.12) $\qquad P(h)\,\Psi_{j,k} = \mu_{j,k}\,\Psi_{j,k} + \mathcal{O}(e^{-\sigma'/h}) \qquad$ for every $\sigma < \sigma' \leqslant S_0 - 2\eta$

If you observe also that :

(4.2.13) $\qquad <\Psi_{j,k} \,/\, \Psi_{j',k'}> = \delta_{(j,k)\,;\,(j',k')} + \mathcal{O}(e^{-\sigma'/h})$

then you can apply proposition 4.1.1 with :

$$A = P(h),\ I = I(h),\ a = a(h),\ \Psi_j = \Psi_{j,k},\ \mu_j = \mu_{j,k}$$

and this gives the estimate.

b) Proof of $\vec{d}(F,E) = \mathcal{O}(e^{-\sigma/h})$

In view of the general properties of the distance between closed spaces recalled in section 4.1, it is sufficient to prove that : $\vec{d}(F,E) < 1$.

Let $u_h$ a normalized eigenvector of $P(h)$ corresponding to one of the $\hat{\lambda}_j(h)$.

Let $\hat{\chi}_j \in C_0^\infty(M_j)$ equal to 1 near $U_j$ and s.t supp $\hat{\chi}_j \cap$ supp $\hat{\chi}_k = \emptyset$ for $j \neq k$. Define $\hat{\chi}_0 \in C^\infty(M)$ such that : $1 = \sum_0^N \hat{\chi}_j$ .

From the results of §3.5, we know that :

(4.2.14) $\qquad \hat{\chi}_0 \cdot u_h = \mathcal{O}(e^{-C_0/h}) \qquad$ in $L^2(M)$ for some $C_0 > 0$.

and that :

$$(4.2.15) \qquad P(h)\,(\hat{\chi}_j\, u_h) = \lambda(h)\,(\hat{\chi}_j\, u_h) + \mathcal{O}(e^{-C_0/h}) \qquad \text{in } L^2(M)$$

Let us prove now that, for every $C_1 < C_0$, we have :

$$(4.2.16) \qquad \hat{\chi}_j\, u_h = \sum_{k=1}^{m_j} a_{jk}\, \varphi_{j,k} + \mathcal{O}(e^{-C_1/h}) \qquad \text{in } L^2(M)$$

with

$$(4.2.17) \qquad \sum_k |a_{jk}|^2 \leq \| \hat{\chi}_j\, u \|^2 \leq 1.$$

The proof of (4.2.16) is immediate for the $h \in \mathfrak{J}$ s.t

$$\| \hat{\chi}_j\, u_h \|_{L^2} \leq e^{-(C_1+C_0)/2h}$$

we just take $a_{jk} \equiv 0$.
Let us look now for the $h \in \mathfrak{J}$ s.t

$$\| \hat{\chi}_j\, u_h \|_{L^2} > e^{-(C_1+C_0)/2h}$$

Then we apply the proposition with $A = P_{M_j}$ and we get, if we denote by $\tilde{E}_j$ the space spanned by the $\varphi_{j,k}$ $(k=1,\ldots,m_j)$ :

$$(4.2.18) \qquad \vec{d}(\mathbb{R}\, \hat{\chi}_j\, u_h, \tilde{E}_j) = \frac{\mathcal{O}(e^{-C_0/h})\,.\,h^{-N_0}}{a(h).\| \hat{\chi}_j\, u_h \|} = \mathcal{O}(h)$$

Then, if we define in this case $a_{jk}$ by :

$$\Pi_{\tilde{E}_j}\, \hat{\chi}_j\, u_h = \sum_k a_{jk}\, \varphi_{jk} \qquad \text{(with (4.2.17) verified)}$$

we get finally (4.2.16) in this second case.
Then, from (4.2.16) and using the decay of $\varphi_{j,k}$ outside $U_j$, we get for some $\varepsilon_0 > 0$ :

$$(4.2.19) \qquad \hat{\chi}_j\, u_h = \sum_{k=1}^{m_j} a_{jk}\, \Psi_{j,k} + \mathcal{O}(e^{-\varepsilon_0/h})$$

Using (4.2.14) and (4.2.19), we finally obtain :

$$(4.2.20) \qquad u_h = \sum_{j=1}^{N} \sum_{k=1}^{m_j} a_{jk}\, \Psi_{j,k} + \mathcal{O}(e^{-\varepsilon_0/h})$$

for some $\varepsilon_0 > 0$ and for $h$ small enough.

Since the $u_h$ form an orthonormal basis for F, we obtain :

$$\vec{d}(F,E) = \mathcal{O}(e^{-C/h}) \qquad \text{for some } C > 0$$

and for h small enough : $\vec{d}(F,E) < 1$.

Remark 4.2.3.

By changing possibly a(h), we see from the proof that we can replace hypothesis (4.2.10) by simply :

$$\operatorname{Sp} P_{M_j}(h) \cap (I(h)+B(0,2a(h))-I(h)) = \emptyset \qquad \text{for } h \in \mathfrak{J}.$$

c) End of the proof.

To prove the last assertion of the proof, let : $\sigma_0 < \sigma < S_0 - 2\eta$ and let us define $a_0 = \exp - \sigma_0/h$.

Let $K_l$ be disjoint intervals s.t

$$I(h) \subset \bigcup_{l \in L} K_l$$

with

$$K_l = ]\alpha_l,\beta_l], \qquad \beta_l - \alpha_l = 2\,a_0.$$

Let $\mathcal{L}$ the set of the l s.t $K_l$ contains one eigenvalue of $P_{M_j}(h)$ or P(h) in I(h). Then the cardinal of $\mathcal{L}$ is $\mathcal{O}(h^{-N_0})$ and we get on this way a covering of $\lambda_1,\dots,\lambda_N$, $\mu_{1,1},\dots,\mu_{N,m_N}$ by intervals $I_1,\dots,I_{\tilde{M}} \subseteq I$ s.t $d(I_j,I_k) \geq 2\,a_0$ if $j \neq k$ and $|I_j| = \mathcal{O}(h^{-N_0})\,a_0$.

Then applying the last assertion of proposition 4.1.1. to A = P for each $I_j$, we get that, in each $I_j$, we have at least as many $\lambda'_s$ as there are $\mu'_s$. But since $N = m_1+\dots+m_N$ from step a) and b) you get exactly the same number. The construction of b is now easy. Because $\sigma_0$ and $\sigma$ are arbitrarily near $S_0 - 2\eta$, the proof of theorem 4.2.1 is complete.

Remark 4.2.4. Proof of Theorem 1.3.1.

Let $\lambda < \varliminf_{|x|\to\infty} V$. Then it is easy to see that we can find a potential W s.t, for $\varepsilon > 0$ chosen s.t $\lambda + \varepsilon < \varliminf_{|x|\to\infty} V$, we have :

$$\text{(4.2.21)} \qquad V^{-1}(\lambda+\varepsilon) = W^{-1}(\lambda+\varepsilon) \overset{\text{dif}}{\equiv} K_\varepsilon$$

$$\text{(4.2.22)} \qquad V = W \text{ on } K_\varepsilon$$

(4.2.23) $\quad W$ satisfies to (1.3.8).

Therefore, decreasing possibly $\varepsilon > 0$, and supposing that $I_\varepsilon = ]\lambda - \frac{\varepsilon}{2}, \lambda + \frac{\varepsilon}{2}[$ is not a critical set for $V = W$, we get that for $\mu \in I_\varepsilon$ :

(4.2.24) $$h^n N_h^W(\mu) - \int_{\xi^2+W(x)<\mu} dx\, d\xi = R_h^W(\mu) = \mathcal{O}(h)$$

where $\mathcal{O}$ is uniform for $\mu \in I_\varepsilon$ .

The only point is to compare now $N_h^V$ and $N_h^W$ . Let us try to apply (a slight generalization of) theorem (4.2.1).

Let us take $I(h) = ]-\infty, \lambda+\varepsilon(h)]$ with

(4.2.25) $$\varepsilon(h) = \mathcal{O}(h^2) \qquad \text{to choose suitably.}$$

Using Theorem 4.2.1 two times with $P^V(h)$ and $P^W(h)$ and choosing the $M_j$ s.t

(4.2.26) $$V = W \text{ on } M_j$$

(4.2.27) $$(V-\lambda)^{-1}(]-\infty, 0]) = \bigcup_j U_j$$

(4.2.28) $$U_j \subset \mathring{M}_j$$

we will get a bijection of the spectrum of $P^V(h)$ in $I(h)$ onto the spectrum of $P^W(h)$ in $I(h)$ if the following condition is satisfied :

(4.2.29) $\quad \exists\, a(h)$ verifying (4.2.5) s.t (4.2.10) is satisfied (see Remark 4.2.3).

Using (4.2.7), it is easy to see that it is always possible to get (4.2.29) for some $\varepsilon_+(h) > 0$ and some $\varepsilon_-(h) < 0$ verifying (4.2.25).

In particular we get :

$$N_h^W(\lambda+\varepsilon_+(h)) = N_h^V(\lambda+\varepsilon_+(h))$$

$$N_h^W(\lambda+\varepsilon_-(h)) = N_h^V(\lambda+\varepsilon_-(h))$$

and then :

(4.2.30) $$N_h^W(\lambda+\varepsilon_-(h)) \leqslant N_h^V(\lambda) \leqslant N_h^W(\lambda+\varepsilon_+(h))$$

We then get easily from (4.2.24) and (4.2.25) that :

(4.2.31) $$h^n N_h^V(\lambda) - \int_{\xi^2+V(x)<\mu} dx\, d\xi = \mathcal{O}(h)$$

#

§4.3 - The tunneling effect.

For $\alpha = (j,k)$, we denote by $j(\alpha)$ the first index. We want to compute the matrix of $P(h)_{\wedge F}$ in a convenient basis. Let us first introduce :

(4.3.1) $$v_\alpha = \pi_F \Psi_\alpha$$

By theorem (4.2.1), we have :

(4.3.2) $$v_\alpha - \Psi_\alpha = \tilde{\mathcal{O}}(e^{-S_0/h}) \qquad \text{in } L^2(M)$$

In the same spirit of remark 4.2.2, we mean that, for each $\varepsilon > 0$, we can find $h_0$ s.t for $0 < h < h_0$, we have for the associate $\Psi_\alpha^{(h)}$ :

$$v_\alpha^{(h)} - \Psi_\alpha^{(h)} = \mathcal{O}(e^{-\frac{S_0}{h} + \frac{\varepsilon}{h}})$$

Let us first remark that the following property :

$$(v_\alpha/v_\beta) = (\Psi_\alpha/\Psi_\beta) - (v_\alpha - \Psi_\alpha/v_\beta - \Psi_\beta)$$

implies, using 4.3.2 :

(4.3.3) $$(v_\alpha/v_\beta) = (\Psi_\alpha/\Psi_\beta) + \tilde{\mathcal{O}}(e^{-2S_0/h})$$

Then, observing that $(\Psi_\alpha/\Psi_\alpha) = 1 + \tilde{\mathcal{O}}(e^{-2S_0/h})$ (see 4.2.8 and 4.2.9), we can write :

(4.3.4) $$S \overset{\text{def}}{\equiv} (\Psi_\alpha/\Psi_\beta) = I + T + \tilde{\mathcal{O}}(e^{-2S_0/h})$$

where $T_{\alpha,\beta} = 0$ if $\alpha = \beta$

$T_{\alpha,\beta} = (\Psi_\alpha/\Psi_\beta)$ if $\alpha \neq \beta$

Let us observe also that :

(4.3.5) $$T = \tilde{\mathcal{O}}(e^{-S_0/h})$$

Indeed, for $j(\alpha) \neq j(\beta)$, this is a simple consequence of the decay of the $\varphi_\alpha$ and if $j(\alpha) = j(\beta)$, $\alpha \neq \beta$, we have more precisely :

$$T_{\alpha,\beta} = \tilde{\mathcal{O}}(e^{-2S_0/h})$$

To compute the matrix of P/F in a suitable basis, we first compute :

$$(4.3.6) \qquad (P\, v_\alpha/v_\beta) = (P\, \Psi_\alpha/\Psi_\beta) - (v_\alpha - \Psi_\alpha \,/\, P(v_\beta - \Psi_\beta))$$

Now, if we define $r_\alpha$ by :

$$(4.3.7) \qquad P\, \Psi_\alpha = \mu_\alpha \Psi_\alpha + r_\alpha$$

we remark that $r_\alpha$ has the same decay properties as $\Psi_\alpha$ , <u>but</u> is supported in the union of small neighborhoods of the $U_k$ (for $k \neq j(\alpha)$), so we get :

$$(4.3.8) \qquad r_\alpha = \tilde{\mathcal{O}}(e^{-S_0/h}) \qquad \text{in } L^2$$

Let us now return to formula (4.3.6). We have :

$$P(v_\beta - \Psi_\beta) = P(\Pi_F - I)\, \Psi_\beta$$

$$= (\Pi_F - I)\, P\, \Psi_\beta = (\Pi_F - I)\, [\mu_\beta \Psi_\beta + r_\beta]$$

$$= \mu_\beta\, (v_\beta - \Psi_\beta) + (\Pi_F - I)\, r_\beta$$

Then it is clear that : $P(v_\beta - \Psi_\beta) = \tilde{\mathcal{O}}(e^{-S_0/h})$ and we get :

$$(4.3.9) \qquad (P\, v_\alpha \,/\, v_\alpha) = (P\, \Psi_\alpha \,/\, \Psi_\beta) + \tilde{\mathcal{O}}(e^{-2S_0/h})$$

We are then reduced, modulo $\tilde{\mathcal{O}}(e^{-2S_0/h})$, to the computation of $(P\, \Psi_\alpha \,/\, \Psi_\beta)$. Using (4.3.7) and the selfadjointness of P(h), we can write :

$$(P\, \Psi_\alpha \,/\, \Psi_\beta) = \frac{\mu_\alpha + \mu_\beta}{2}\, (\Psi_\alpha \,/\Psi_\beta) + 1/2[(r_\alpha \,/\, \Psi_\beta) + (\Psi_\alpha \,/\, r_\beta)]$$

$$((P\, \Psi_\alpha \,/\, \Psi_\beta)) = \text{diag}\, \mu_\alpha + \frac{1}{2}\, T(\text{diag}\, \mu_\alpha) + \frac{1}{2}\, (\text{diag}\, \mu_\alpha).T$$

$$+ \frac{1}{2}\, [((\Psi_\alpha \,/\, r_\beta)) + ((r_\alpha \,/\, \Psi_\beta)) + \tilde{\mathcal{O}}(e^{-2S_0/h})]$$

If we now remember that

$$r_\alpha = [P, \chi_{j(\alpha)}]\ \varphi_\alpha$$

then we get :

$$(\Psi_\alpha / r_\beta) = h^2 \int \chi_{j(\alpha)} (\varphi_\beta \cdot \nabla \varphi_\alpha - \varphi_\alpha \nabla \varphi_\beta) \nabla \chi_{j(\beta)}\, dx$$

$$+ h^2(\nabla \chi_{j(\alpha)} \varphi_\alpha | \nabla \chi_{j(\beta)} \varphi_\beta)$$

$$= h^2 \int \chi_{j(\alpha)}(\varphi_\beta \cdot \nabla \varphi_\alpha - \varphi_\alpha \nabla \varphi_\beta) \nabla \chi_{j(\beta)}\, dx + \tilde{\mathcal{O}}(e^{-2S_0/h})$$

So we get finally :

$$(4.3.10) \qquad ((P\, v_\alpha / v_\beta)) = \operatorname{diag} \mu_\alpha + \frac{1}{2} T.\operatorname{diag} \mu_\alpha + \frac{1}{2} \operatorname{diag} \mu_\alpha T + (\hat{W}_{\alpha,\beta}) + \tilde{\mathcal{O}}(e^{-2S_0/h})$$

with

$$(4.3.11) \qquad \hat{W}_{\alpha,\beta} = \frac{1}{2} (W_{\alpha,\beta} + W_{\beta,\alpha})$$

$$(4.3.12) \qquad W_{\alpha,\beta} = h^2 \int \chi_{j(\alpha)} (\varphi_\beta \cdot \nabla \varphi_\alpha - \varphi_\alpha \nabla \varphi_\beta) \nabla \chi_{j(\beta)}\, dx$$

It is better in fact to compute the matrix of P/F in an orthonormal basis deduced from the basis $(v_\alpha)$ by the classical procedure :

$$(4.3.13) \qquad e_\alpha = \sum_\beta v_\beta\, (V^{-1/2})_{\beta,\alpha}$$

Then using (4.3.3) and (4.3.4), we deduce from (4.3.10) the following :

<u>Theorem 4.3.1.</u>

The matrix of $P_{/F}$ in the basis $(e_\alpha)$ is given by

$$(4.3.14) \qquad \operatorname{diag}(\mu_\alpha) + \hat{W}_{\alpha,\beta} + \tilde{\mathcal{O}}(e^{-2S_0/h})$$

where $\hat{W}_{\alpha,\beta}$ is defined in (4.3.11), (4.3.12).

<u>Remark 4.3.2.</u>

It is not difficult to see, using the decay of the $\varphi_\alpha$ , that

$$(4.3.15) \qquad \hat{W}_{\alpha,\beta} = \tilde{\mathcal{O}}(e^{-S_0/h})$$

we then recover Theorem (4.2.1). Indeed if you compare the spectra of 2 self adjoint matrices of size $\mathcal{O}(h^{-N_0})$ whose difference is $\tilde{\mathcal{O}}(e^{-S_0/h})$ (or $\tilde{\mathcal{O}}(e^{-2S_0/h})$), then, there exists a bijection b between their two spectra s.t

$$b(\lambda) - \lambda = \tilde{\mathcal{O}}(e^{-S_0/h}) \qquad \text{(resp. } \tilde{\mathcal{O}}(e^{-2S_0/h}))$$

Then, the spectrum of $\mathrm{diag}(\mu_\alpha) + \hat{W}_{\alpha,\beta}$ determines the spectrum of P(h) in I(h) modulo $\tilde{\mathcal{O}}(e^{-2S_0/h})$.

#

Remark 4.3.3. (Return to the one-well case)
The proof of theorem (4.3.1) gives that when you pass from P(h) to a Dirichlet Problem $P_{M_1}(h)$ in some open set $M_1$ containing $B(U_1,S)$ then there exists a bijection between $\mathrm{Sp}\, P_{M_1} \cap I(h)$ and $\mathrm{Sp}\, P(h) \cap I(h)$ s.t $b(\lambda) - \lambda = \tilde{\mathcal{O}}(e^{-2S/h})$

This proves lemma 3.4.3.

#

Formula (4.3.14) introduces the matrix $\hat{W}$ which corresponds to the interaction between the different wells. In fact, we can observe from (4.3.12), that

$$W_{\alpha,\beta} = \tilde{\mathcal{O}}(e^{-2S_0/h}) \qquad \text{for } j(\alpha) = j(\beta) \tag{4.3.16}$$

We will now try to compute $\hat{W}_{\alpha,\beta}$ for $j(\alpha) \neq j(\beta)$ in a way which permits to study the effect of the interaction.
Let us first remark that if we have, say two wells $U_1, U_2$ and if we have only one eigenvalue $\mu_1(\mu_2)$ attached to each well, then if

$$|\mu_1 - \mu_2| \geq e^{-\varepsilon_0/h} \qquad \text{for some } \varepsilon_0 > 0 \quad (\varepsilon_0 < S_0)$$

the tunneling effect does'nt change the basic properties and we can say that the two wells are independent (non resonant) (see [HE-SJ]$_3$ for a general study of these properties). P(h) admits in this case 2 eigenvalues $\tilde{\mu}_1$ and $\tilde{\mu}_2$ which are separate from each other by at least $e^{-\varepsilon_0/2h}$ and the corresponding eigenfunctions are located (modulo exponentially small error) respectively in the neighborhood of $U_1$ and $U_2$. So the problem of tunneling appears in this situation if $\mu_1 - \mu_2$ is exponentially small for order $e^{-S_0/h}$.

This is the reason which leads us to introduce two new hypotheses

(4.3.17) $\mu_\alpha - \mu_\beta = \mathcal{O}(h^\infty)$

(4.3.18) $\varphi_\alpha = \mathcal{O}(h^{-N_0} e^{-d(x,U_{j(\alpha)})/h})$

We have given in section 3.3 sufficient conditions to get (4.3.18).
The verification of (4.3.17) is obtained frequently by symmetry hypotheses between different wells which implies : $\mu_\alpha = \mu_\beta$ (see[HE-SJ]$_2$)

---

Let us observe now that by an integration by part in the formula (4.3.12) we get :

(4.3.19) $W_{\alpha,\beta} - W_{\beta,\alpha} = (\mu_\alpha - \mu_\beta)\, t_{\alpha,\beta}$

We then get that under the hypotheses(4.3.17) and (4.3.18) that :

(4.3.20) $\begin{cases} \text{The matrix of } P/F \text{ in this basis } (e_\alpha) \text{ is given by :} \\ \operatorname{diag}(\mu_\alpha) + W_{\alpha,\beta} + \mathcal{O}(h^\infty)\, e^{-S_0/h} \end{cases}$

We now try to compute $W_{\alpha,\beta}$ mod $\mathcal{O}(h^\infty)\, e^{-S_0/h}$ (for $j(\alpha) \neq j(\beta)$) with the hope that the principal contribution in the study of the spectrum of $\operatorname{diag}(\mu_\alpha) + W_{\alpha,\beta}$ will not be perturbed too much by the error $\mathcal{O}(h^\infty)\, e^{-S_0/h}$.

We mean by that, that if the splitting between two eigenvalues we want to determine is of order $h^{-\nu}.e^{-S_0/h}$ then the perturbation $\mathcal{O}(h^\infty)\, e^{-S_0/h}$ does'nt affect the result.

Until the end of the subsection, we write $\equiv$ for : equal modulo $\mathcal{O}(h^\infty)\, e^{-S_0/h}$ and we suppose that (4.3.17) and (4.3.18) are satisfied.
Then, let us first remark that :

(4.3.21) $W_{\alpha,\beta} \equiv 0$ if $d(U_{j(\alpha)}, U_{j(\beta)}) > S_0$
or
if $j(\alpha) = j(\beta)$

Let us now consider the case when :

(4.3.22) $j(\alpha) \neq j(\beta)$ ; $d(U_{j(\alpha)}, U_{j(\beta)}) = S_0$

and let us consider more carefully the integral (4.3.12) defining $W_{\alpha,\beta}$ in this case.

To take account of the fact that the integration for $x \in M$ s.t

$$d(x,U_{j(\alpha)}) + d(x,U_{j(\beta)}) > S_0 + a \qquad \text{(with } a > 0\text{)}$$

gives a contribution $\equiv 0$, we introduce :

$$E^{(a)} = \{x \in M,\ d(U_{j(\alpha)},x) + d(U_{j(\beta)},x) \leqslant S_0 + a\} \tag{4.3.23}$$

It is clear that, for $a > 0$ (and $\eta > 0$) sufficiently small, we have the inclusion :

$$E^{(a)} \subset M_{j(\alpha)} \cup M_{j(\beta)} \tag{4.3.24}$$

and that $E^{(a)}$ does'nt contain other wells.
Then we can always find some open set $\Omega$ with smooth boundary s.t

$$U_{j(\alpha)} \subset \Omega\ ,\quad U_{j(\beta)} \cap \overline{\Omega} = \emptyset$$

$$E^{(a)} \cap \overline{\Omega} \subset M_{j(\alpha)}\ ,\ E^{(a)} \cap \complement\Omega \subset M_{j(\beta)}$$

and then $\Gamma = \partial\ \Omega \cap E^{(a)}$ is compact in $M_{j(\alpha)} \cap M_{j(\beta)}$

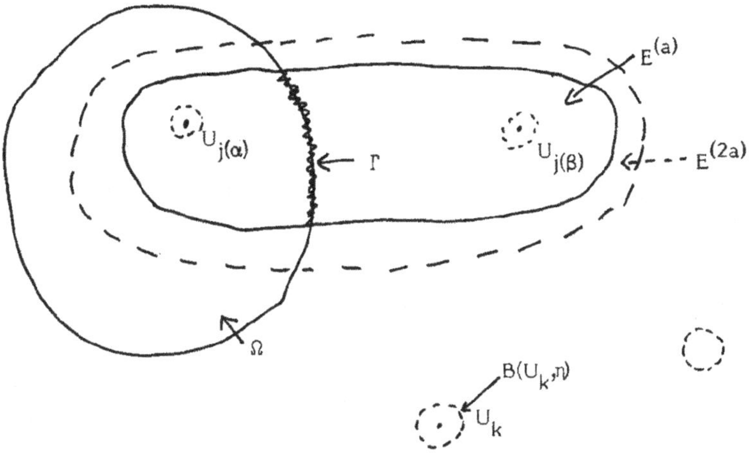


We choose a sufficiently small s.t $E^{(2a)}$ has the same properties. Let us introduce now $\chi_{E^{(a)}} \in C_0^\infty(M)$ equal to 1 in $E^{(a)}$ and with support in $E^{(2a)}$. Then we have :

$$\operatorname{supp}\ \chi_{E^{(a)}} \cap \overline{\Omega} \subset M_{j(\alpha)}$$

$$\operatorname{supp}\ \chi_{E^{(a)}} \cap \complement\Omega \subset M_{j(\beta)}$$

and so $\chi_{j(\alpha)}$ (resp. $\chi_{j(\beta)}$) is equal to 1 on supp $\chi_E(a) \cap \bar{\Omega}$ (resp. supp $\chi_E(a) \cap \complement \Omega$)

Let us now compute $W_{\alpha,\beta}$ .

$$W_{\alpha,\beta} \equiv h^2 \int \chi_E(a) \cdot \chi_{j(\alpha)}(\varphi_\alpha \cdot \nabla \varphi_\beta - \varphi_\beta \cdot \nabla \varphi_\alpha) \nabla \chi_{j(\beta)} \, dx$$

$$\equiv h^2 \int_\Omega \chi_E(a) (\varphi_\alpha \nabla \varphi_\beta - \varphi_\beta \nabla \varphi_\alpha) \nabla \chi_{j(\beta)} \, dx$$

and using the Green formula :

$$(4.3.25) \qquad W_{\alpha,\beta} = h^2 \int_\Omega \nabla^* (\chi_E(a) (\varphi_\alpha \nabla \varphi_\beta - \varphi_\beta \nabla \varphi_\alpha) \cdot \chi_{j(\beta)} \, dx$$

$$+ h^2 \int_{\partial\Omega} \langle \chi_E(a) (\varphi_\alpha \nabla \varphi_\beta - \varphi_\beta \nabla \varphi_\alpha), \frac{\partial}{\partial n} \rangle \chi_{j(\beta)} \, d S_{\partial\Omega}$$

where $\frac{\partial}{\partial n}$ is the exterior normal derivative to $\partial\Omega$

and $d S_{\partial\Omega}$ is the induced measure on $\partial\Omega$.

Using (4.3.17) and (4.3.18) and the property that the $\varphi_\alpha$ are eigenfunctions of $P_{M_{j(\alpha)}}$,

we see that the first term in the right hand side of (4.3.25) is $\equiv$ to 0. We get then finally :

$$(4.3.26) \qquad W_{\alpha,\beta} \equiv h^2 \int_\Gamma (\varphi_\alpha \frac{\partial \varphi_\beta}{\partial n} - \varphi_\beta \frac{\partial \varphi_\alpha}{\partial n}) \, d S_\Gamma$$

So we have proved the :

<u>Theorem 4.3.4.</u>

Under the hypotheses(4.3.17) and (4.3.18) and the hypotheses of theorem 4.3.1 (see §4.2), the matrix of $P_{\wedge F}$ in the basis $e_\alpha$ is given by :

$$(4.3.27) \qquad \operatorname{diag}(\mu_\alpha) + \tilde{W}_{\alpha,\beta} + \mathcal{O}(h^\infty) e^{-S_0/h}$$

with

$$(4.3.28) \quad \begin{cases} \tilde{W}_{\alpha,\beta} = 0 & \text{if } d(U_{j(\alpha)}, U_{j(\beta)}) > S_0 \text{ or } j(\alpha) = j(\beta) \\[2ex] \tilde{W}_{\alpha,\beta} - h^2 \int_\Gamma (\varphi_\alpha \cdot \frac{\partial \varphi_\beta}{\partial n} - \varphi_\beta \cdot \frac{\partial \varphi_\alpha}{\partial n}) \, d S_\Gamma & \text{if } d(U_{j(\alpha)}, U_{j(\beta)}) = S_0 \end{cases}$$

Let us remark that this new theorem is unsatisfactory because we don't know for the moment about the precise behavior of the $\varphi_\alpha$ but we can see from the formula (4.3.28) that we need only the knowledge of $\varphi_\alpha$ in a neighborhood of $\Gamma$. We will see in the next subsection how to exploit this idea.

Application 4.3.5 the double well problem.

Suppose now that we are in the case of two wells and that in $I(h)$ : $m_1 = m_2 = 1$. We denote simply the eigenvalue of $P_{M_j}$ by $\mu_j$ (j=1,2).

Then we have :

$$S_0 = d(U_1,U_2)$$

$$\beta \overset{\text{déf}}{\equiv} \mathscr{W}_{1,2} = \mathscr{W}_{2,1} = \mathcal{O}(h^{-N_0})\, e^{-S_0/h}$$

Then we get from theorem (4.3.4), that the matrix of P/F is simply :

$$(4.3.29)\qquad \begin{pmatrix} \mu_1 & 0 \\ 0 & \mu_2 \end{pmatrix} + \begin{pmatrix} 0 & \beta \\ \beta & 0 \end{pmatrix} + \mathcal{O}(h^\infty)\, e^{-S_0/h}$$

Then you get from (4.3.29) that P(h) has two eigenvalues $\lambda_1$ , $\lambda_2$ s.t :

$$(4.3.30)\qquad |\lambda_2-\lambda_1| = \sqrt{4\,\beta^2+(\mu_1-\mu_2)^2} + \mathcal{O}(h^\infty)\, e^{-S_0/h}$$

This information is significative if for example :

$$(4.3.31)\qquad \sqrt{4\,\beta^2+(\mu_1-\mu_2)^2} > C\, h^\nu\, e^{-S_0/h} \qquad \text{for some } \nu \in \mathbb{R} \text{ and some } C > 0$$

We have explained (after 4.3.16) the phenomenon in the case when $|\mu_1-\mu_2| \geqslant C\, h^\nu\, e^{-S_0/h}$ ($\nu$ negative large enough).

Let us now consider the case when there exists an isometry g of M s.t $g^2 = e$ and $g.U_1 = U_2$. Then using the symmetry properties (see $[\text{HE-SJ}]_2$ for the general situations) we get that the matrix (4.3.29) has the form :

$$(4.3.32)\qquad \mu\, I_2 + \begin{pmatrix} 0 & \tilde\beta \\ \tilde\beta & 0 \end{pmatrix}$$

with $\tilde\beta = \beta + \mathcal{O}(h^\infty)\, e^{-S_0/h}$

If $\beta$ verifies $|\beta| \geqslant C\, h^\nu\, e^{-S_0/h}$, we get the splitting between $\lambda_1$ and $\lambda_2$ :

$$(4.3.33)\qquad |\lambda_2-\lambda_1| = \beta\, e^{-S_0/h} + \mathcal{O}(h^\infty)\, e^{-S_0/h} \geqslant C'\, h^\nu\, e^{-S_0/h}$$

More precisely, in the basis $(e_1,e_2)$, the normalized eigenvectors $u_1$ , $u_2$ are given by

$$\begin{pmatrix} 1/\sqrt{2} \\ 1/\sqrt{2} \end{pmatrix} \quad \text{and} \quad \begin{pmatrix} 1/\sqrt{2} \\ -1/\sqrt{2} \end{pmatrix}$$

In particular, you can see that

$u_1+u_2$ is localized near $U_1$

and that

$u_1-u_2$ is localized near $U_2$

and that the probability to be near $U_1$ for $u_1$ is equal to the probability for $u_2$ to be near $U_2$.

Remark 4.3.6.

Starting from a symmetric situation, it is interesting to study the effect of a small perturbation outside the wells (called by B. Simon $[SI]_5$ the flea on the elephant). Our approach is particularly adapted to the study of this problem (see $[HE\text{-}SJ]_2$, [J.M.S.]).

## §4.4 Study of the multiple wells problem at the bottom.

In the preceeding subsection, Theorem 4.3.4. reduces (modulo $\mathcal{O}(h^{\infty})\, e^{-S_0/h}$) the problem of the interaction between the different wells to the computation of $\mathcal{W}_{\alpha,\beta}$ (formula 4.3.28). This formula is rather implicit because we need to know the behavior of $\varphi_\alpha$. In this section, we will use the B.K.W. construction realized at the bottom to get more explicit results.

We make in this section the following assumptions :

(4.4.1) $E = 0 = \min V \quad (< \lim_{|x|\to\infty} V$ in the case $M = \mathbb{R}^n)$

(4.4.2) $V^{-1}(0) = \bigcup_{j=1}^{N} U_j$ (disjoint union)

(4.4.3) $U_j = \{x_j\}$ and $V''(x_j) > 0$ for $j=1,\dots,N$

and we define : $S_0 = \inf_{j\neq k} d(x_j,x_k)$.

First of all let us introduce our notion of geodesics. Let $\tilde{M}$ be the manifold M equipped with the Agmon Metric $V\,dx^2$. By $\tilde{M}$-geodesic, we mean a continuous curve $[a,b]\ni t\to x(t)\in M$ which up to reparametrization is a geodesic outside the $\{x_j\}$. In this section, we omit now the reference to $\tilde{M}$ and use always the Agmon distance.

Definition 4.4.1.

For $1 \leq j \leq N$, let $\Omega_j$ consist of $x_j$ and the union of the interiors of all minimal geodesics from $x_j$ to some point in M, of length strictly less than : $S_0 = \min_{j\neq k} d(x_j,x_k)$.

Here we consider only geodesics $\gamma : [0,a] \ni t \to M$ s.t

$$\gamma(t) \in M \setminus \{x_1,\dots,x_N\} \quad \text{for } t \in ]0,a]$$

and

$$\gamma(t) \xrightarrow[t \to 0]{} x_j$$

As it is classical (see [AB,MA]), a geodesic of the form $\gamma$ is a space-projection of an integral curve of the Hamiltonian field $\frac{\xi^2}{V}$.

After reparametrization, we may assume that this integral curve is in $\frac{\xi^2}{V} = 1$ and after another parametrization, we may consider $\gamma$ as the projection of a null-bicaracteristic of $q = \xi^2 - V(x)$ (but the parametrization is now on a semi-infinite interval $]-\infty,\tilde{b}]$).

The following proposition is more or less well known in Geometry (at least in the case where the metric is non-degenerate) (see [AB-MA], [MI] and $[\text{HE-SJ}]_1$ p. 404. Prop. 6.5).

Proposition 4.4.2.

$\Omega_j$ has the following properties :

(4.4.4) $\Omega_j$ is an open set

(4.4.5) $d_j(x) = d(x,U_j)$ is $C^\infty$ on $\Omega_j$

(4.4.6) Let $y \in \Omega_j$ and let $]-\infty,0] \ni t \to x(t)$ the integral curve of $\nabla d_j$ s.t. $x(0) = y$. Then $x(-\infty) = \{x_j\}$

(4.4.7) For each $y \in \Omega_j$ there exists a unique geodesic joining $x_j$ and $y$ whose length is $d_j(y)$ and which is given by (4.4.6).

As a corollary, we see that we can extend the B.K.W. construction given in a neighborhood of $\{x_j\}$ in section (2.3) (Theorem 2.3.1) and relative to the first eigenvalue of the harmonic oscillator associated to the well $\{x_j\}$. Indeed, we have no problem to define the phase in $\Omega_j$ which is simply $d_j(x)$. To define the amplitude you have just to integrate along the integral curves of $\nabla d_j$ starting from the inital values on a small sphere $d_j = \mathfrak{J}_0 > 0$ (let us just recall that all the transport equations were of the type

$$(X + b)\, u = f \quad \text{with} \quad X = 2\, \nabla d_j \quad \text{(see 2.3.6, 2.3.9, etc))}$$

So we get the :

Proposition 4.4.3. (see Theorem 2.3.1)

Under Hypotheses (4.4.1) - (4.4.3), we can find a formal series

(4.4.8) $$E^{(j)}(h) \sim \sum_{k=1}^{\infty} E_k^j h^k$$

and a formal symbol defined in $\Omega_j$

(4.4.9) $$a^{(j)}(x,h) \sim \left( \sum_{k=0}^{\infty} a_k^{(j)}(x)\, h^k \right)$$

s.t with $\theta_j(x,h) = h^{-n/4} \,.\, a^{(j)}(x,h)\, e^{-d_j(x)/h}$

(4.4.10) $$(P(h) - E^{(j)}(h)) \quad \theta_j = \mathcal{O}(h^{\infty})\, e^{-d_j(x)/h} \qquad \text{in } \Omega_j$$

Moreover

(4.4.11) $$a_0^{(j)}(x) > 0 \qquad \text{in } \Omega_j$$

The property (4.4.11) is just a consequence of the fact that, if y was a point of $\Omega_j$ with $a_0^{(j)}(y) = 0$, this would be true along the integral curve of $\nabla d_j$ ; so, in particular, we would get $a_{(0)}^{(j)}(x_j) = 0$ which is in contradiction with (2.3.2).

We can in fact suppose that

(4.4.12) $$\| \theta_j \|_{L^2(\Omega_j)} = 1$$

Let us look now at the formula giving $\mathcal{W}_{\alpha,\beta}$ . We take now :

(4.4.13) $I(h) = [0,Ch]$ where C is chosen s.t for some $N^+ \in \mathbb{N}$ we have :

1) for $j = 1,\dots,N^+$, $P_{M_j}$ has exactly one eigenvalue in I(h) and, with $a(h) = \varepsilon_0 h$ ($\varepsilon_0$ small enough), the condition $\mathrm{Sp}\, P_{M_{j(h)}} \cap (I(h) + B(0,2a(h)) - I(h)) = \emptyset$ is satisfied.

2) for $j = N^+ +1,\dots,N$, $P_{M_j}$ has no eigenvalues in $I(h) + B(0,2a(h))$.

Observe that it is always possible to be in this situation and that, using remark (4.2.3), (4.2.10) is then satisfied.

In this situation, the $\{x_j\}$ for $j = N^+ +1,\dots,N$ are called the non resonnant wells.

The $\{x_j\}$ for $j = 1,\dots,N_+$ are called the resonnant wells.

Because $m_j = 1$, we just denote now $\tilde{\mathcal{W}}_{\alpha,\beta}$ by $\tilde{w}_{j,k}$ and we have just to compute : $\tilde{w}_{j,k}$ for $j \neq k$ s.t $d(x_j,x_k) = S_0$ . (j and k belong only to $[1,\dots,N^+]$).

Our idea will be now to replace in formula (4.3.28) the Dirichlet eigenfunction $\varphi_j$ of $P_{M_j}$ associated to $\mu_j$ by the corresponding B.K.W. solution $\Theta_j$. For this, we need only to compare $\varphi_j$ and $\Theta_j$ in a neighborhood of the set $\Gamma_{j,k}$ introduced in section (4.3) which is clearly inside $\Omega_j$.

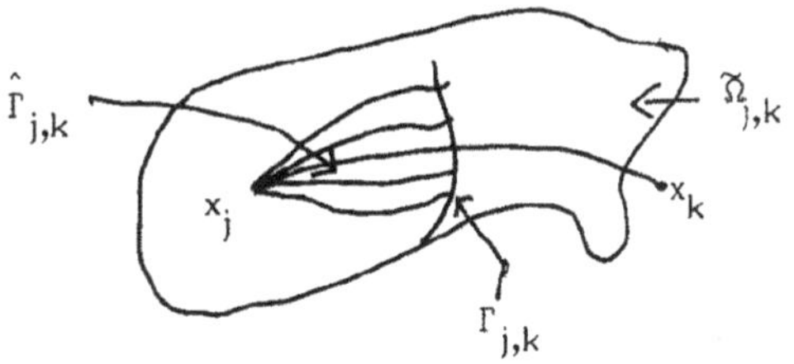


Let $\hat{\Gamma}_{j,k}$ the set described by the minimal geodesics joining $\Gamma_{j,k}$ to $\{x_j\}$. And let $\tilde{\Omega}_{j,k}$ ($\subset \Omega_j$) be a small neighborhood of $\hat{\Gamma}_{j,k}$ with smooth boundary. We want to prove the following :

Theorem 4.4.4.

If $\varphi_j$ is the eigenfunction of $P_{M_j}(h)$ with eigenvalue $\mu_j$ in $I(h)$, then, if $\tilde{\chi}_j$ is a $C^\infty$ function equal to 1 in a neighborhood of $\hat{\chi}_{j,k}$ and supported in $\tilde{\Omega}_{j,k}$, we have :

$$(4.4.14) \qquad w \equiv \tilde{\chi}_j(\varphi_j - \Theta_j) = O(h^\infty)\, e^{-d_j(x)/h} \qquad \text{in } H^2(\tilde{\Omega}'_{j,k})$$

where $\tilde{\Omega}'_{j,k}$ is some neighborhood of $\hat{\Gamma}_{j,k}$.

Proof

As usual (see the proof of proposition 3.3.4), we shall first prove (4.4.14) in $H^1(\tilde{\Omega}'_{j,k})$ and then by using the standard regularity theorem for the Dirichlet Problem in $\tilde{\Omega}'_{j,k}$ we will get the theorem in $H^l(\tilde{\Omega}'_{j,k})$ (for $l \geq 2$).

In spirit, this proof is related to the proofs of Carleman estimates.
More recently, J. Sjöstrand [SJ] makes an intensive use of these technics for the study of analytic singularities. But the proof here is in fact more elementary, due to the particularity of the Schrödinger equation.
In what follows, we note $\Omega = \tilde{\Omega}_{j,k}$ (to simplify the notations).
Let us start from the basic identity (3.1.1) in $\Omega$:

$$(4.4.15) \qquad h^2 \| \nabla (e^{\tilde{\Phi}/h} w) \|^2_{\Omega} + \int_{\Omega} ((V-\mu_j) - |\nabla \tilde{\Phi}|^2)\, w^2 e^{2\tilde{\Phi}/h} dx = \int_{\Omega} e^{2\tilde{\Phi}/h} ((P-\mu_j)w).w.dx.$$

Let us first observe that $\tilde{\chi}_j \ \theta_j$ is an approximate eigenfunction of $P_{M_j}$ (with $\tilde{\tilde{\chi}}_j = 1$ on supp $\tilde{\chi}_j$), so by the comparison theorem we have :

(4.4.16) $$w \equiv \mathcal{O}(h^{\infty}) \quad \text{in } \Omega_j$$

We have also, because $\varphi_j$ and $\theta_j$ have this property

(4.4.17) $$w = \mathcal{O}(h^{-N_0})\, e^{-d_j(x)/h}$$

Let us observe now that :

$$(P-\mu_j)w = \tilde{\chi}_j(P-\mu_j)(\varphi_j - \theta_j) + [P,\tilde{\chi}_j](\varphi_j - \theta_j)$$

and we get :

(4.4.18) $$(P-\mu_j)w = \mathcal{O}(h^{\infty})\, e^{-d_j(x)/h} + \tilde{r}$$

where $\tilde{r}$ satisfies to :

(4.4.19) $$\tilde{r} = \mathcal{O}(h^{-N_0})\, e^{-d_j(x)/h} \qquad \text{in } \tilde{\Omega}_j \text{ ; supp } \tilde{r} \subset \text{supp } \nabla\tilde{\chi}_j$$

We want to prove that :

(4.4.20) $$w = \mathcal{O}_N(h^{N})\, e^{-d_j(x)/h}, \quad \forall N \in \mathbb{N} \text{ in a neighborhood of } \hat{\Gamma}_{j,k}$$

To use conveniently (4.4.15), we need to find a phase $\tilde{\Phi}_N$ which has the following properties, for some $N_0$, independent of N :

(4.4.21) $$(P-\mu_j)w = h^{-N_0}\, \mathcal{O}(e^{-\tilde{\Phi}_N(x)/h}) \qquad \text{in } \Omega$$

(4.4.22) There exists a decomposition of $\Omega = \Omega^{+} \cup \Omega^{-}$ s.t

(1) $$w = h^{-N_0}\, \mathcal{O}(e^{-\tilde{\Phi}_N(x)/h}) \qquad \text{in } \Omega^{-}$$

(2) $$(V-\mu_j(h) - |\nabla \tilde{\Phi}_N|^2) \geq C_0\, h^{N_0} \qquad \text{in } \Omega^{+} \qquad (\text{with } C_0 > 0)$$

(4.4.23) $$\tilde{\Phi}_N(x) \geq d_j(x) + N\, h \,\text{Log}\,(1/h) - N_0\, h\, \text{Log}\, 1/h \qquad \text{in a neighborhood } \hat{\Omega} \text{ of } \hat{\Gamma}_{j,k}.$$

Under these conditions, it is clear, following the proof of proposition 3.3.1, that you get

$$(4.4.24)\qquad w = h^{-3N_0}\, \mathfrak{S}_N(e^{-\tilde{\Phi}_N(x)/h}) \qquad \text{in } \Omega$$

and, in particular, due to the property $(4.4.22)_2$, (4.4.20).
Let us take :

(4.4.25)

(1) $\Omega^-(h) = \{x \in M,\ d_j(x) \leq C\, h\}$

(2) $\Omega^+(h) = \Omega \setminus \Omega^-(h)$

Let us observe that (4.4.18) suggests to take $\tilde{\Phi}_N$ as the minimum of two phases because we have the two conditions (with $N_1$ independent of N).

(4.4.26)

(1) $\tilde{\Phi}_N(x) \leq d_j(x) + N\, h \operatorname{Log} 1/h + N_1\, h \operatorname{Log} 1/h$

(2) $\tilde{\Phi}_N(x) \leq d_j(x) + N_1\, h \operatorname{Log} 1/h$ on the support of $\nabla \tilde{\chi}_j$

All these considerations lead to the following choice of $\tilde{\Phi}_N$ inspirated of course by the proof of proposition 3.3.5 (which was only sketched).
Let us take (we omit j) :

$$(4.4.27)\qquad \tilde{\Phi}_N(x) = \min(\phi(x) + N\, h \operatorname{Log} 1/h,\ \Psi(x))$$

where

(4.2.28)

(1) $\phi(x) = d(x) - C\, h \operatorname{Log}\left(\frac{d(x)}{h}\right)$ in $\Omega^+$

(2) $\phi(x) = d(x) - C\, h \operatorname{Log} C$ in $\Omega^-$

$$(4.4.29)\qquad \Psi(x) = \inf_{y \in \operatorname{supp} \nabla\tilde{\chi}} [\phi(y) + (1-\varepsilon) d(x,y)]$$

Then let us verify (4.4.21) to (4.4.26).
(4.4.26) (and (4.4.21) through (4.4.18)) is satisfied for $N_0$ large enough (depending of C) if we observe that $\Psi(x) = \phi(x)$ on supp $\nabla \tilde{\chi}$.
$(4.4.22)_1$ is satisfied in $\Omega^-$ as a consequence of $(4.4.25)_1$ and (4.4.16)
$(4.4.22)_2$ is satisfied if C is well chosen. Indeed we have :

$$|\nabla \Psi|^2 \leq (1-\varepsilon)^2\, V \qquad \text{a.e}$$

and then

(4.4.30) $\quad V(x) - |\nabla \Psi|^2 \geqslant (2\varepsilon-\varepsilon^2)\, V(x)$

and

$$V(x) - |\nabla \phi|^2 = 0 \qquad \text{in } \Omega^-$$

$$\overset{\text{a.e}}{=} V(x) - |\nabla d|^2 \left(1 - \frac{Ch}{d}\right)^2 \overset{\text{a.e}}{\geqslant} \frac{V(x)}{d(x)}\, Ch$$

If we note the existence of $C_0 > 0$ s.t $\frac{V(x)}{d(x)} \geqslant C_0$ in $\Omega$ we get :

(4.4.31) $\quad V(x) - |\nabla \phi|^2 \geqslant \frac{C}{C_0}\, h \qquad$ a.e in $\Omega^+$

If we choose now $\frac{C}{C_0} \geqslant E_1 + 1$ where $E_1$ is the first eigenvalue of the harmonic oscillator associated to $\{x_j\}$, we get $(4.4.22)_2$ by observing that in small ball $B(x_j, \mathfrak{I}_0)$ containing $\{x_j\}$ we have

(4.4.32) $\quad \phi(x) + N\, h \operatorname{Log} 1/h \leqslant \Psi(x) \qquad$ in $B(x_j, \mathfrak{I}_0)$

In fact, we have the following lemma :

Lemma 4.4.5.

There is a sufficiently small neighborhood $\hat{\Omega}$ of $\hat{\Gamma}_{j,k}$ and $\varepsilon > 0$ s.t :

(4.4.33) $\quad d_j(x) \leqslant (1-\varepsilon)\,[d_j(y)+d(y,x)] \qquad \forall\, y \in \operatorname{supp} \nabla \tilde{\chi}_j \qquad \forall x \in \hat{\Omega}$

(4.4.33) is the consequence of the fact that we can choose $\hat{\Omega}$ s.t the minimal geodesic joining $\hat{\Omega}$ to $\{x_j\}$ avoid a neighborhood of supp $\nabla\tilde{\chi}_j$ .

#

Then choosing this $\varepsilon$ in the definition of $\Psi$ it is easy to verify (4.4.32) for $h \in \,]0,h_N]$ and in $\hat{\Omega}$, so we get

(4.4.34) $\quad \tilde{\Phi}_N(x) = \phi(x) + Nh \operatorname{Log} \frac{1}{h} \qquad$ in $\hat{\Omega}$

which gives the property (4.4.23) for convenient $N_0$ .
This finishes the proof of proposition 4.4.4.

#

<u>The computation of</u> $\tilde{w}_{j,k}$

We can now return to the computation of $\tilde{w}_{j,k}$. Recall the formula given in (4.3.28) :

(4.4.35) $$w_{j,k} \equiv h^2 \int_{\Gamma_{j,k}} (\varphi_j \frac{\partial \varphi}{\partial n} - \varphi_k \frac{\partial \varphi_j}{\partial n}) \, d S_\Gamma \qquad (\mathrm{mod}\ \mathcal{O}(h^\infty)\, e^{-S_0/h})$$

Using theorem (4.4.4), we get :

$$w_{j,k} \equiv h^2 \int_{\Gamma_{j,k}} ( \theta_j \frac{\partial \theta_k}{\partial n} - \theta_k \cdot \frac{\partial \theta_j}{\partial n}) \, d S_{\Gamma_{j,k}}$$

and, using proposition (4.4.3), we obtain finally :

(4.4.436) $$w_{j,k} \equiv -h^{1-n/2} \int_{\Gamma_{j,k}} [a^{(j)} a^{(k)} (\frac{\partial d_k}{\partial n} - \frac{\partial d_j}{\partial n}) + h(a^{(j)} \frac{\partial a^{(k)}}{\partial n} - a^{(k)} \frac{\partial a^{(j)}}{\partial n} ]\times$$
$$\times\, e^{-\frac{d_j(x)+d_k(x)}{h}} \, d S_{\Gamma_{j,k}}$$

Decreasing $\Gamma_{j,k}$ if necessary (and transversal to the minimal geodesics joining $\Gamma_{j,k}$ to $x_j$ and $x_k$) we observe that $\frac{\partial d_k}{\partial n}(x)$ and $-\frac{\partial d_j}{\partial n}(x)$ are different of 0 and not very different of $V(x)$. Then, for h sufficiently small, we have $a^{(j)} a^{(k)} (\frac{\partial d_k}{\partial n} - \frac{\partial d_j}{\partial n}) \geq C > 0$ on $\Gamma_{j,k}$

We then get the inequality

(4.4.37) $$C \int_{\Gamma_{j,k}} e^{-\frac{d_j(x)+d_k(x)}{h}} \, d S_{\Gamma_{j,k}} \leq -h^{n/2-1} w_{j,k} \leq \frac{1}{C} \int_{\Gamma_{j,k}} e^{-\frac{d_j(x)+d_k(x)}{h}} \, d S_{\Gamma_{j,k}}$$

for $C > 0$ sufficiently small

(4.4.37) implies first the inequality

(4.4.38) $$-h^{n/2-1} w_{j,k} \leq C e^{-S_0/h}$$ because $d_j(x) + d_k(x) \geq S_0$

To prove the contrary let $x_{j,k}$ one point in $\mathring{\Gamma}_{j,k}$ s.t

$$d(x_j, x_{j,k}) + d(x_{j,k}, x_k) = S_0$$

and observe that :

$$d_j(x) + d_k(x) \leqslant S_0 + C\, d(x,x_{j,k})^2 \qquad \text{on } \Gamma_{j,k}$$

Then we get :

(4.4.39) $$C\, e^{-S_0/h}\, h^{n-1/2} \leqslant -h^{n/2-1}\, w_{j,k}$$

Finally we get the :

Theorem 4.4.6. ([HE-SJ]$_1$)

Let I(h) as defined in (4.4.13), then for $j \in [1,\dots,N_+]$, $k \in [1,\dots,N_+]$ s.t $d(x_j,x_k) = S_0$, there exists a constant $C > 0$ s.t

(4.4.40) $$\frac{1}{C}\, h^{1/2} \leqslant -\tilde{w}_{j,k}\, e^{S_0/h} \leqslant C\, h^{1-n/2}$$

Remark 4.4.7.

This theorem can be empty if there are no wells $x_j, x_k$ with $j \in [1,\dots,N_+]$, $k \in [1,\dots,N_+]$ s.t $d(x_j,x_k) = S_0$. More precise results in this case are given in [HE-SJ]$_3$.

To improve theorem (4.4.6), we add now the two following geometrical hypotheses

(4.4.41) For $j,k \in [1,\dots,N_+]$ s.t $d(U_j,U_k) = S_0$ ,

there is a finite number of geodesics $\gamma^{(l)}_{j,k}$ ($l \in \Lambda_{j,k}$) of length $S_0$ joining $x_j$ and $x_k$. #

(4.4.42) Let $x^{(l)}_{j,k} \in \gamma^{(l)}_{j,k} \cap \Omega_j \cap \Omega_k$ and let $\Gamma^{(l)}_{j,k} \subset\subset \Omega_j \cap \Omega_k$

be a smooth hypersurface which intersects $\gamma^{(l)}_{j,k}$ transversally at $x^{(l)}_{j,k}$ and s.t $x^{(l)}_{j,k}$ is the only point in $\overline{\Gamma}^{(l)}_{j,k} \cap \gamma^{(l)}_{j,k}$ . Then there exists a constant $C > 0$ s.t

$$d_{j,k}(x) \overset{\text{déf}}{=} d(x,x_j) + d(x,x_k) \geqslant S_0 + \frac{1}{C}\, d(x,x^{(l)}_{j,k})^2 \qquad \text{for } x \in \Gamma^{(l)}_{j,k} \qquad \#$$

Then, using the stationnary phase theorem, we get the

Theorem 4.4.8. ([HE-SJ]$_1$)

Let I(h) as defined in (4.4.13) and suppose that (4.4.41) and (4.4.42) are satisfied. Then for $j \in [1,\dots,N_+]$, $k \in [1,\dots,N_+]$ s.t $d(x_j,x_k) = S_0$ we have :

(4.4.43) $$\tilde{w}_{j,k} \equiv -h^{1/2} \left( \sum_{m=0}^{\infty} b^m_{j,k}\, h^m \right) e^{-S_0/h}$$

with

$$b^0_{j,k} = 2 \sum_{l \in \Lambda_{j,k}} (2\Pi)^{-(\frac{n-1}{2})} \frac{a_0^{(j)}(x^{(l)}_{j,k})a_0^{(k)}(x^{(l)}_{j,k})\, V^{1/2}(x^{(l)}_{j,k})}{[\text{Hess}^{tr}\, d_{jk}(x^{(l)}_{jk})]^{1/2}}$$

Remark 4.4.9.

Theorem (4.4.6) is an improvement of a theorem of B. Simon $[SI]_3$, which says only that $\lim_{h \to 0} 1/h \operatorname{Log}(-\tilde{w}_{j,k}) = -S_0$

Remark 4.4.10.

The problem consisting in finding analogs of (4.4.40) and (4.4.43) for higher levels is studied by Martinez [MA] and also in $[\text{HE-SJ}]_2$.

Exercice 4.4.11.

Explicit the results in the application (4.3.5) (see section 4.5 for the case n = 1).

§4.5 - The case n = 1.

Let us apply the general results to the most simple case which motivates the study and for which the best results where known before (see [RE-SI] Vol. IV, [LA-LI], E. Harrell [HA], S. Coleman [CO], ...).

Let us consider the potential : $V(x) = 1/4\,(x^2-1)^2$

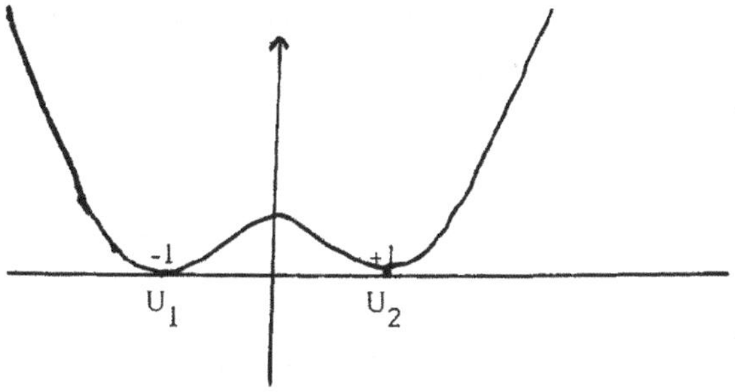


Taking E = 0, we have two wells

(4.5.1) $\qquad V^{-1}(0) = U_1 \cup U_2 \qquad$ with $U_1 = \{-1\}$ , $U_2 = \{+1\}$

Observe also the symmetry :

(4.5.2) $\qquad V(-x) = V(x)$

Let us just recall the different steps of the general proof in this particular case.

Step 1 The localized harmonic oscillators.

Because $V''(-1) = V''(+1) = 2$, the localized harmonic oscillator at $U_1$ (resp $U_2$) is given in the new coordinate $y = x + 1$ (resp. $y = x - 1$) by :

(4.5.3) $$-h^2 \frac{d^2}{dy^2} + y^2$$

whose first eigenvalue is h.

From now on, we look for the spectrum in :

(4.5.4) $$[0,2h]$$

Step 2 The decay of the eigenfunctions.

The Agmon distance is given by :

(4.5.5) $$d(x,y) = \int_x^y |(t^2-1)|\, dt \qquad \text{if } x < y.$$

The eigenfunctions $u_h(x)$ of $-h^2 \frac{d^2}{dx^2} + V(x)$ associated to eigenvalues have a priori the following decay (see Prop. 3.3.3, 3.5.3).

Case $x < -1$ :

$$\forall M,\ u_h(x) = \mathcal{O}_M(h^{-N_0})\, e^{-1/h \int_x^{-1} \sqrt{V(t)}\, dt}$$

uniformly for $x \in ]-M,-1]$ (and the same for derivatives)

$$\forall \varepsilon > 0 \quad u_h(x) = \mathcal{O}_\varepsilon(e^{\varepsilon/h})\, e^{-(1-\varepsilon)/h \int_x^{-1} \sqrt{V(t)}\, dt}$$

for $x \in ]-\infty,-1[$ (in $H^1$ $(]-\infty,-1])$)

Case $-1 < x \leq 0$ :

$$u_h(x) = \mathcal{O}(h^{-N_0})\, e^{-1/h \int_{-1}^{x} \sqrt{V(t)}\, dt} \qquad \forall x \in ]-1,0]$$

Case $0 \leq x \leq 1$ :

$$u_h(x) = \mathcal{O}(h^{-N_0})\, e^{-1/h \int_x^{1} \sqrt{V(t)}\, dt}$$

Case $x > 1$ :

See the case $x < -1$ and use the symmetry.

Step 3 Introduce Dirichlet Problems.

As was explained in the general context, we introduce 2 Dirichlet problems

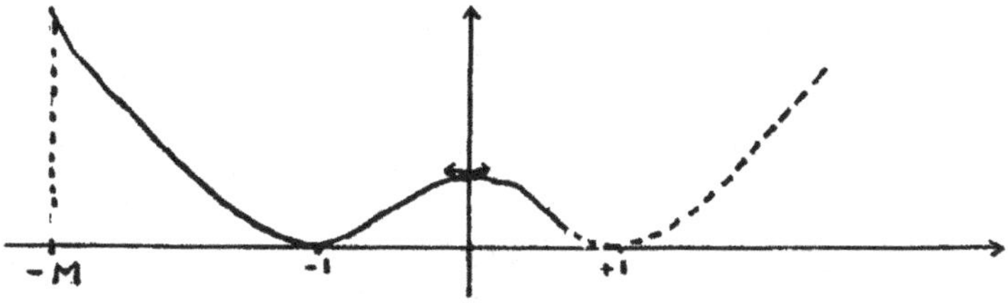


in $M_1 = ]-M, 1-\eta[$

and symmetrically in

$$M_2 = ]-1+\eta, M[$$

$\eta$ is chosen arbitrarily small ($\eta > 0$) and M is chosen s.t

$$\int_{-M}^{-1} \sqrt{V(t)}\ dt \geqslant 2\, S_0$$

where :

(4.5.6) $$S_0 = \int_{-1}^{+1} \sqrt{V(t)}\ dt = 5/6$$

is the Agmon distance between the two wells.

Then, in I(h), there exists only one eigenvalue of the Dirichlet problem $P_{M_1}(h)$ : $\mu_1(h)$ which admits an expansion of the form

(4.5.7) $$\mu_1(h) = h + \sum_{j \geqslant 2} E_j h^j$$

By symmetry there exists in I(h), only one eigenvalue of the Dirichlet problem $P_{M_2}(h)$ : $\mu_2(h)$ and we have :

(4.5.8) $$\mu_2(h) = \mu_1(h)$$

If $\varphi_1(h)(x)$ is a normalized eigenfunction associated to $\mu_1(h)$, it is clear that $\varphi_2(h)(x) \overset{\text{déf}}{\equiv} \varphi_1(h)(-x)$ is a normalized eigenfunction for $P_{M_2}(h)$ associated to $\mu_2(h) = \mu_1(h)$. Taking this choice of eigenfunctions we then have :

(4.5.9) $$\varphi_1(h)(x) = \varphi_2(h)(-x), \qquad \forall\, x \in M_1$$

and we know from §3.3 that :

$$(4.5.10)\qquad \varphi_1(h)(x) = \mathcal{O}(h^{-N_0})\, e^{-1/h \int_{-1}^{x} \sqrt{V(t)}\, dt}$$

uniformly for $x \in ]-1,1-\eta[$ and the same for the derivatives.

Step 4 Compute the interaction matrix in an adapted basis.

Let us recall how we get this special basis. First of all, we introduce

$$(4.5.11)\qquad \Psi_1(h)(x) = \chi_1(x)\, \varphi_1(h)(x)$$

where $\begin{cases} \chi_1(x) = 1 \text{ in } ]-M+\eta, 1-3/2\,\eta[ \\ \operatorname{supp} \chi_1 \subset ]-M, 1-\eta[ \end{cases}$

and we define $\Psi_2$ by symmetry :

$$(4.5.12)\qquad \Psi_2(h)(x) = \Psi_1(h)(-x) \qquad \text{for } x \in \mathbb{R}$$

Then if F is the spectral space of P(h) associated to I(h), we introduce

$$(4.5.13)\qquad w_i = \Pi_F\, \Psi_i \qquad i = 1,2$$

$(w_1, w_2)$ is an (almost orthonormal) basis of the 2-dimensional space F and we orthonormalize this basis by introducing :

$$(4.5.14)\qquad \vec{e} = \vec{w} \,.\, S^{-1/2} \qquad \text{with } S = (w_i / w_j)$$

In this new basis, and, because we have respected at each step the symmetry of the problem, the matrix of P/F in the basis $\vec{e}$ is given by :

$$(4.5.15)\qquad \mathcal{M} = \begin{pmatrix} \tilde{\mu} & \tilde{\beta} \\ \tilde{\beta} & \tilde{\mu} \end{pmatrix}$$

with

$$(4.5.16)\qquad \begin{cases} \tilde{\mu}(h) = \mu_1(h) + \mathcal{O}(e^{-2S_0/h} \,.\, e^{2\eta/h}) \\ \tilde{\beta}(h) = \beta(h) + \mathcal{O}(h^{\infty})\, e^{-S_0/h} \\ \beta(h) = -h^2(\varphi_1^{(h)}(0)\varphi'^{(h)}_2(0) - \varphi'^{(h)}_1(0).\varphi_2^{(h)}(0)) \end{cases}$$

Then, see (4.3.32), we get that $\mathcal{M}$ has two eigenvalues $\lambda_1(h), \lambda_2(h)$ with $\lambda_1(h) \leq \lambda_2(h)$ s.t

$$(4.5.17)\quad \begin{aligned}\lambda_2(h) - \lambda_1(h) &= -2\ \tilde{\beta}(h)\\ &= 2\ h^2[\varphi_1^{(h)}(0)\ \varphi'^{(h)}_2(0) - \varphi'^{(h)}_1(0)\ \varphi_2^{(h)}(0)] + \mathcal{O}(h^\infty)\ e^{-S_0/h}\end{aligned}$$

(here we have admitted that $\tilde{\beta} \neq 0$, see the computation we will make after).

<u>Step 5</u> <u>Computation of</u> $\varphi_1^{(h)}(0), \varphi'^{(h)}_1(0)$.

We just use the B.K.W. approximation. We have seen in §4.4 that :

$$(4.5.18)\quad \varphi_1^{(h)}(x) = h^{-1/4}\ \pi^{-1/4}\ a(x,h)\ e^{-1/h \int_{-1}^{x} \sqrt{V(t)}\,dt} \qquad \text{for } x \in [-1, 1-2\eta[$$

where

$$(4.5.19)\quad a(x,h) \sim \sum_{j=0}^{\infty} a_j(x)\ h^j$$

Then $a_0(x)$ can be computed by solving the transport equation :

$$(4.5.20)\quad a_0(x) = e^{-\int_{-1}^{x} \frac{(V^{1/2})'(t)-1}{2\sqrt{V(t)}}\,dt}$$

From (4.5.17) and (4.5.20), we get

$$(4.5.21)\quad \lambda_2(h) - \lambda_1(h) = 4\ h^{1/2}\ \pi^{-1/2}\ e^{-A}\ [1 + \mathcal{O}(h)]\ e^{-S_0/h}$$

where

$$(4.5.22)\quad A = \lim_{\varepsilon \to 0} \Big[ \int_{-1+\varepsilon}^{1-\varepsilon} \frac{dx}{2\sqrt{V(x)}} + \operatorname{Log} \varepsilon \Big]$$

This formula is given (in a slightly different form) by Harrell [H A] and there is a different heuristic proof of Coleman [CO] using the method of Instantons. In the spirit of this last paper, we have the following interpretation for A. Let

$$q(x,\xi) = \xi^2 - V$$

q describes the motion in the forbidden region (we have replaced V by -V)

graph of -V

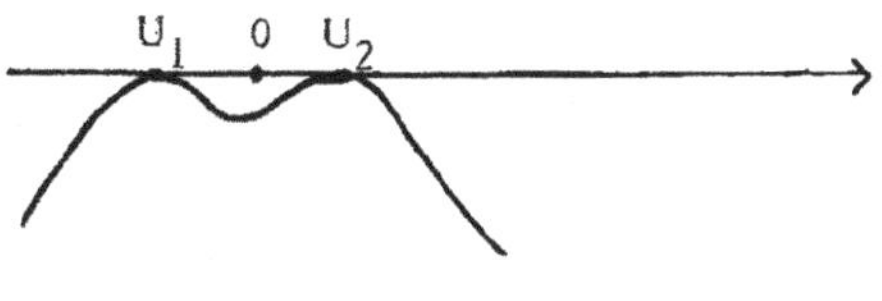

and we just look at the motion between the two tops of the mountain.
The motion equation gives :

$$\frac{dx}{dt} = 2\,\xi = 2\,\sqrt{V(x(t))}$$

and we get : $$dt = \frac{dx}{2\,\sqrt{V(x)}}$$

$\int_{-1+\varepsilon}^{1-\varepsilon} \frac{dx}{2\,\sqrt{V(x)}}$ is the time $T(\varepsilon)$ you need to go from $(-1+\varepsilon)$ to $(1-\varepsilon)$

$T(\varepsilon)$ tends to $\infty$ like $-\mathrm{Log}\,\varepsilon$ when $\varepsilon \to 0$ and so A is the " principal part " of $T(\varepsilon)$.

## §5 AN INTRODUCTION TO RECENT RESULTS OF WITTEN.

### §5.1 - Introduction

The purpose of this section is to present results suggested by Witten [WI] as an application of the results presented in §4. We want just to explain how Witten get the Morse Inequalities from semi-classical Analysis. We must first refer to two nice expositions of these results by Henniart in Bourbaki Seminar [HEN] and more recently in the book [C.F.K.S] where you can find a complete elementary proof of the Morse Inequalities in the Spirit of Witten. In fact, you don't need all the strength of the results we have presented before, because the approximation by the localized Harmonic oscillators is sufficient. However, we will present other results about the Witten-Laplacian in which you need a part of the more precised theory developped in §2.4 and we refer to $[HE\text{-}SJ]_4$ for deeper results concerning the Witten complex. Let us finally say that these proofs " à la Witten " are not the simplest ones but I think they give a new bridge between Analysis and Geometry.

Let M be a compact $C^\infty$ Riemannian oriented n-dimensional Manifold and let $f : M \to \mathbb{R}$ be a $C^\infty$ Morse function that means a $C^\infty$ function s.t all the critical points are non degenerate.

At each critical point $U_j$ $(j=1,\dots,N)$, we attach the index $l_j$ corresponding to the number of (-) in the signature of the Hessian of f at the point $U_j$ : $f''(U_j)$.

For $l = 0,\dots,n$, we denote by :

$$(5.1.1) \qquad \mathcal{C}^{(l)} = \{j\ ;\ U_j \text{ is of index } l\}$$

and we define $m_l$ as :

$$(5.1.2) \qquad m_l = \text{Card } \mathcal{C}^{(l)}$$

Let us now recall one definition (the definition for the Analysts ! ) of the Betti numbers. Let d be the differential on M and let us denote by $\Lambda^p(M)$ the space of the $C^\infty$ p-forms. Because M is Riemannian, we have a natural scalar product defined on $\Lambda^p(M)$, and, taking the completion for the associated norm, we can define (because M is compact) the Hilbert space of the $L^2$ p-forms $\Lambda^p_{L^2}(M)$.

The restriction of d to $\Lambda^p(M)$ is denoted by $d_p$ :

$$(5.1.3) \qquad \Lambda^p(M) \xrightarrow{d_p} \Lambda^{p+1}(M)$$

and we denote by $d_p^*$ the formal adjoint :

$$(5.1.4) \qquad \Lambda^{p+1}(M) \xrightarrow{d_p^*} \Lambda^p(M)$$

$d^*$ is the differential operator whose restriction to $\Lambda^p(M)$ is $d^*_{p-1}$ .

On the space $\Lambda(M)$ (= $\bigoplus_{p=0}^{n} \Lambda^p(M)$), the following operator is well defined as a differential operator :

(5.1.5) $\quad - \Delta = (d+d^*)^2 = dd^* + d^* d \qquad$ (recall that $d^2 = 0$)

By restriction to the p-forms, you get the so-called Laplace-Beltrami operator on the p-forms :

(5.1.6) $\quad - \Delta_p = d_p^* \, d_p + d_{p-1} \cdot d_{p-1}^*$

On $\Lambda^p(M)$ (see [C-F-K-S] for a self contained exposition), $-\Delta_p$ is essentially selfadjoint and admits consequently a unique self-adjoint extension as an unbounded operator on $\Lambda^p_{L^2}(M)$.
The principal symbol of $-\Delta_p$ is at a point $(x,\xi)$ of $T^* M \setminus \{0\}$ given by :

(5.1.7) $\quad (\sum_{i,j} g_{ij} (x) \xi_i \xi_j) I_{\mathcal{L}(\Lambda^p T_x^* M)} \qquad$ (see 1.2.3)

The operator is consequently elliptic of order 2 and we know that in this case the Resolvent is compact. In particular, the Kernel of $-\Delta_p$ is finite dimensional, and we define the $p^{th}$ Betti number $b_p$ by :

(5.1.8) $\quad b_p = \dim \mathrm{Ker}\,(-\Delta_p)$

Another equivalent way to define the Betti numbers is to consider the de Rham complex :

(5.1.9) $\quad 0 \to \Lambda^0(M) \xrightarrow{d_0} \Lambda^1(M) \xrightarrow{d_1} \qquad \to \Lambda^n(M) \to 0$

(let us recall that $d_{p+1} \circ d_p = 0$ as a consequence of $d^2 = 0$)
and you get $b_p$ as :

(5.1.10) $\quad b_p = \dim(\mathrm{Ker}\, d_{p/\mathrm{Range}(d_{p-1})})$

Indeed, the Hodge theory gives a natural bijection between $\mathrm{Ker}\, d_{p/\mathrm{Range}(d_{p-1})}$ and $\mathrm{Ker}(-\Delta_p)$.

The Morse Inequalities can be written in different forms. First of all, the weak Morse Inequalities (W.M.I) say :

(5.1.11) $\quad$ (W.M.I.): $m_l \geq b_l \quad$ for $l = 0,\ldots,n$.

The strong Morse Inequalities (S.M.I) can be written in the following way. Let us introduce the polynômials :

(5.1.12) $P(t) = \sum_{i \geq 0} b_i t^i$ ; $M(t) = \sum_{i \geq 0} m_i t^i$

Then the strong Morse Inequalites say :

(5.1.13) (S.M.I.)There exists a polynomial Q with coefficients in $\mathbb{Z}^+$ s.t $M(t) - P(t) = (1+t)\,Q(t)$

In particular, if $t = -1$, we get :

(5.1.14) $$\sum_{i \geq 0} (-1)^i m_i = \sum_{i \geq 0} (-1)^i b_i$$

The number : $\sum_{i \geq 0} (-1)^i b_i$ is the well known Euler Characteristic $\chi(M)$.

An equivalent way to formulate (S.M.I) is to say that there exists a complex $\mathring{E}$ of vector spaces over $\mathbb{R}$

(5.1.15) $$0 \to E^0 \xrightarrow{u_0} E^1 \to \qquad \xrightarrow{u_n} E^n \to 0$$

with $\dim_{\mathbb{R}} E^k = m_k$

$$\dim_{\mathbb{R}} H^k(E) \overset{\text{déf}}{\equiv} \dim(\mathrm{Ker}\, u_k / \mathrm{Range}\, u_{k-1}) = b_k$$

The proof of this equivalence is just elementary algebra.

Following Witten, we will try to find such a complex of finite dimensional vector spaces having the same cohomology as the de Rham Complex. For this purpose, we introduce a perturbation of the de Rham Complex now called the Witten Complex depending on a small parameter $h \in ]0,h_0]$ :

(5.1.16) $d_f = e^{-f/h}(h\, d)\, e^{f/h}$ (where f is the Morse function)

and it is clear that the new complex has the same cohomology as the de Rham complex :

(5.1.17) $$\dim(\mathrm{Ker}\, d_f^{(p)} / \mathrm{Range}(d_f^{(p-1)})) = b_p$$

The Hodge theory works in this new context. We introduce the Witten's Laplacian :

(5.1.18) $$\mathcal{P}_f(h) = (d_f + d_f^*)^2$$

which gives by restriction on the p-forms :

(5.1.19) $\mathcal{P}_f^{(p)}(h) = d_f^{(p)*}\, d_f^{(p)} + d_f^{(p-1)}\, d_f^{*(p-1)}$

For each $h > 0$, $\mathcal{P}_f^{(p)}(h)$ has the same properties as $-\Delta^{(p)}$.
In particular, for $A > 0$ fixed, let us consider :

(5.1.20) $\begin{cases} E_f^{(p)}(A,h), \text{ the eigenspace of } \mathcal{P}_f^{(p)}(h) \text{ associated to the eigenvalues belonging} \\ \text{to } [0, A\, h^{3/2}] \end{cases}$

From the Hodge theory, we get :

(5.1.21) $b_p = \dim \operatorname{Ker} \mathcal{P}_f^{(p)}(h)$

and we have the trivial inclusion :

(5.1.22) $\operatorname{Ker} \mathcal{P}_f^{(p)}(h) \subset E_f^{(p)}(A,h)$

and consequently :

(5.1.23) $b_p \leq \dim E_f^{(p)}(A,h) \qquad \forall A \in \mathbb{R}^+, \quad \forall h \in ]0,h_0]$

The proof of the W.M.I. is now reduced to the proof of :

(5.1.24) $\exists A > 0$ and $h \in ]0,h_0]$ s.t $\qquad \dim E_f^{(p)}(A,h) = m_p$

In fact we shall prove in the next subsection that it is true for each $A > 0$ and for each $h \in ]0,h_0(A)]$.

---

Let us now remark that we have :

(5.1.25) $d_f\, \mathcal{P}_f(h) = \mathcal{P}_f(h)\, d_f$

and consequently

(5.1.26) $d_f^{(p)}\, E_f^{(p)}(A,h) \subset E_f^{(p+1)}(A,h)$

and (by taking the adjoint in 5.1.25)

(5.1.27) $d_f^{(p)*} E_f^{(p+1)}(A,h) \subset E_f^{(p)}(A,h)$

Then $d_f \wedge \mathring{E}_f(A,h)$ defines a complex of finite dimensional space whose cohomology is the same as the de Rham Cohomology.

(5.1.28) According to (5.1.15), (5.1.24) gives also the proof of the (S.M.I).

//

§5.2. - Proof of the Strong Morse Inequalities.

Let us compute the Witten's Laplacian $(d_f+d^*_f)^2 = \mathcal{T}_f(h)$. We get :

(5.2.1) $\tilde{\partial} = hd \quad , \quad d_f = \tilde{\partial} + df \wedge$

(5.2.2) $d^*_f = \tilde{\partial}^* + \nabla f \lrcorner$

(where $\nabla f \lrcorner$ is the interior product).

Then :

$$(d_f+d_{f*})^2 = d_f.d^*_f+d^*_f.d_f = (\tilde{\partial}+df\wedge)\,(\tilde{\partial}^*+\nabla f \lrcorner)+(\tilde{\partial}^*+\nabla f \lrcorner)(\tilde{\partial}+df\wedge)$$

$$= (\tilde{\partial}+\tilde{\partial}^*)^2 + ((df\wedge).\nabla f \lrcorner) + (\nabla f \lrcorner).(df\wedge) + \tilde{\partial}.(\nabla f \lrcorner) + (\nabla f \lrcorner).\tilde{\partial} + (df\wedge).\tilde{\partial}^* + \tilde{\partial}^*.(df\wedge)$$

Then we use the relations :

$$(df\wedge)\,(\nabla f \lrcorner) + (\nabla f \lrcorner)\,(df\wedge) = |\nabla f|^2 \,.\, I$$

$$\mathcal{L}_{\nabla f} = d(\nabla f \lrcorner) + (\nabla f \lrcorner).d$$

to get

(5.2.3) $(d_f+d_{f*})^2 = -h^2\,\Delta + |\nabla f|^2 + h(\mathcal{L}_{\nabla f} + \mathcal{L}^*_{\nabla f})$

Then, it is not to difficult, to prove the following lemma :

Lemma 5.2.1.

$\mathcal{L}^*_{\nabla f} + \mathcal{L}_{\nabla f}$ is a $C^\infty(M)$. linear operator whose matrix at a critical point $U_j$ is given on the p-forms by :

(5.2.4) $C_f^{(p)}(U_j) = 2\, f''(U_j)^{(p)} - \mathrm{Tr}\, f''(U_j)\,.\, I$

where $f''(U_j)^{(0)} = 0$

$f''(U_j)^{(1)}$ is the Hessian of f at $U_j$ that we consider here as an application in $\mathcal{L}(T^*_{U_j}M, T^*_{U_j}M)$

and $f''(U_j)^{(p)}$ is defined for any $(u_1,\dots,u_p)$ in $T^*_{U_j}M$ by

(5.2.5) $$f''(U_j)^{(p)}(u_1 \wedge \dots \wedge u_p) = (f''(U_j)u_1) \wedge u_2 \wedge \dots \wedge u_p + u_1 \wedge (f''(U_j)u_2) \wedge \dots \wedge u_p + \dots + (u_1 \wedge u_2 \dots \wedge (f''(U_j)u_p))$$

By taking on orthonormal basis $e^j_i$ for $f''(U_j)$ corresponding to the eigenvalues $\lambda^j_i$ $(i=1,..,n)$, which we arrange s.t :

(5.2.6) $$\lambda^j_1 \leq \lambda^j_2 \dots \leq \lambda^j_{\ell_j} < 0 < \lambda^j_{\ell_{j+1}} \leq \dots \leq \lambda^j_n$$

(we have supposed that f is a Morse function so $f''(U_j)$ is not degenerate) it is not difficult to verify that the eigenvalues of $f''(U_j)^{(p)}$ are given by :

$$\lambda^j_{i_1} + \dots + \lambda^j_{i_p} \qquad \text{for } i_1 < i_2 < \dots < i_p$$

and that the corresponding eigenvectors are :

$$e^j_{i_1} \wedge e^j_{i_2} \wedge \dots \wedge e^j_{i_p} \; .$$

So we get finally :

<u>Lemma</u> <u>5.2.2.</u>

The eigenvalues of $C_f^{(p)}(U_j)$ are given by :

(5.2.7) $$2(\lambda^j_{i_1} + \dots + \lambda^j_{i_p}) - \sum_{i=1}^{n} \lambda^j_i$$

for any $I = (i_1,\dots,i_n)$ s.t $i_1 < i_2 < \dots < i_p$

and where the $\lambda^j_i$ are the eigenvalues of $f''(U_j)$.

Let us now follow the general strategy of the previous sections to analyse the spectrum of $\mathcal{P}_f^{(p)}(h)$ in the interval : $I(h) = [0, Ah^{3/2}]$ for A large enough. To apply this strategy to this (a little) more general situation, we just need to have a weak result on the decay of the eigenfunctions outside the wells (The wells here are just the critical points of f ! ).

As in §3, where all the decay results are deduced from theorem 3.1.1, all the results on the decay of eigenfunctions can be obtained from the following theorem.

Theorem 5.2.3.

Let $\Omega$ be a bounded open set with $C^\infty$ boundary in M ; let $u \in C^2(\overline{\Omega}, \Lambda^P)$ with $u_{\wedge_{\partial\Omega}} = 0$ and let $\phi$ be a Lipschitzian function on $\overline{\Omega}$, then we have :

$$\operatorname{Re}(e^{2\phi/h}\, \mathcal{P}_f^{(p)}(h)u,u)$$

(5.2.8)
$$= \| \tilde{d}^*(e^{\phi/h}u)\|^2 + \|\tilde{d}(e^{\phi/h}u)\|^2$$

$$+ ((|\nabla f|^2 - |\nabla\varphi|^2 + h\, C_f^{(p)})\, e^{\phi/h}u \,/\, e^{\phi/h}u)$$

Proof.

As in §3, we reduce, by a regularization argument, to the case when $\phi \in C^2(\overline{\Omega}, \mathbb{R})$. Then we can use the following formula which is also useful for the construction of B.K.W. solutions.

(5.2.9)
$$(e^{\phi/h}\, \mathcal{P}_f^{(p)}(h)\, e^{-\phi/h}) = -h^2 \Delta^{(p)} + |\nabla f|^2 - |\nabla\phi|^2 + h(\mathcal{L}_{\nabla f} + \mathcal{L}^*_{\nabla f} + \mathcal{L}_{\nabla\varphi} - \mathcal{L}^*_{\nabla\phi})$$

#

We have now all the material to prove the (S.M.I). Using weak analogs of Theorem 4.2.1 and of theorem (3.4.1) we get a bijection b between the spectrum of the direct sum of the " harmonic oscillators systems " attached to each well $U_j$ : $Q_j^{(p)}(h)$ and the spectrum. of $\mathcal{P}_f^{(p)}(h)$ in I(h) s.t

(5.2.10)
$$|b(\lambda) - \lambda| < C\, h^{3/2}$$

Let us define more precisely this system of harmonic oscillators.

Take normal coordinates x in the neighborhoof of $U_j$ (that means in particular that the matrix of the metric at $U_j$ is the identity).

Then the localized system at $U_j$ is given in these coordinates by :

(5.2.11)
$$Q_j^{(p)}(h) = [-h^2 \sum_{i=1}^{n} \frac{d^2}{dx_i^2} + (f''(U_j)\, x \,/\, f''(U_j)x)]\, I + h\, C_f^{(p)}(U_j)$$

and let us observe (to get 5.2.10) that :

(5.2.12)
$$\|(\mathcal{P}_f^{(p)}(h) - Q_j^{(p)}(h))\, u\| \leq C\, h^{3/2} \|u\|_3$$

for $u \in \Lambda^P(\Omega_j)$ with compact support in a small open neighborhood $\Omega_j$ of $U_j$.

Now if we choose $A \geqslant C$ and h sufficiently small, we get that :

(5.2.13) $\left\{\begin{array}{l}\text{the number of eigenvalues of } \mathcal{S}_f^{(p)}(h) \text{ belonging to } [0,Ah^{3/2}] \text{ is exactly the} \\ \text{numbers of zero eigenvalues of } \bigoplus_j Q_j^{(p)}(1).\end{array}\right.$

Let us look at the eigenvalues of $Q_j^{(p)}(1)$. Using (5.2.7) and (5.2.11), we obtain :

(5.2.14) $\left\{\begin{array}{l}\text{The eigenvalues of } Q_j^{(p)}(1) \text{ are of the form :} \\ \sum_{i=1}^{n} (2\alpha_i+1)\, |\lambda_i^j| + 2(\lambda_{i_1}^j + \dots + \lambda_{i_p}^j) - \sum_{i=1}^{n} \lambda_i^j \\ \text{where } \alpha \in (\mathbb{Z}^+)^n \text{ and } I = (i_1,\dots,i_p) \text{ satisfies } 1 \leqslant i_1 < i_2 < \dots < i_p \leqslant n\end{array}\right.$

Now we can rewrite (5.2.14) using the index $l_j$ (see 5.2.6) :

(5.2.15) $$\sum_{i=1}^{n} 2\alpha_i\, |\lambda_i^j| + 2 \sum_{i=1}^{l_j} |\lambda_i^j| + 2(\lambda_{i_1}^j + \dots + \lambda_{i_p}^j)$$

Then we see that the eigenvalues are always positive (which was necessary because it corresponds to eigenvalues of the first approximation of a positive operator ! ) and that the only zero eigenvalue (if it exists) corresponds to the case : $p = l_j$ , $i_1 = 1, \dots, i_p = l_j$. So we get :

(5.2.16) $\left\{\begin{array}{l}\text{If } p \neq l_j \text{ , Ker } Q_j^{(p)}(1) = \{0\} \\ \text{If } p = l_j \text{ , Ker } Q_j^{(p)}(1) = \mathbb{R}.e^{-\varphi_0(x)}\, e_1^j \wedge \dots \wedge e_p^j\end{array}\right.$

where $\varphi_0(x)$ is the phase associated to the quadratic potential :

$$x \to (f''(U_j)\, x \,/\, f''(U_j)\, x)$$

Finally we get from (5.2.13) and (5.2.16) that :

(5.2.17) $\left\{\begin{array}{l}\text{the number of eigenvalue of } \mathcal{S}_f^{(p)}(h) \text{ belonging to } [0,Ah^{3/2}] \text{ is, for A large,} \\ \text{enough and } h < h_A \text{, equal to } m_p \text{ .}\end{array}\right.$

According to the arguments of section 5.1 (see 5.1.24, 5.1.28), this ends the proof of the Strong Morse Inequalities.

Exercise 5.2.4. (See [WI], [HE-SJ]$_4$)

Starting from formula (5.2.9) and using the Agmon distance $\phi(x) = d(x,U_j)$, find the B.K.W. solution associated to the well $U_j$ (in the case $l_j = p$) in the neighborhood of $U_j$.

It is not the purpose of these lectures to present all the results obtained in [HE-SJ]$_4$ (which need more precise estimates than the estimates we have given in §3 and 4) to measure the " Instantons " effects and justify completely the Witten's programm [WI] on the Morse functions. We also don't discuss other aspects of Witten's paper relative to Euler-Poincaré theorem and to Hirzebruch signature theorem (see [ELS-WA]). Let us mention also many interesting problems relative to the degenerate case (see Bismut [BI] and [HE-SJ]$_{5,6}$). We just want to explain how to get the following :

Proposition 5.2.5.

The eigenvalues in $[0,Ah^{3/2}]$ of $\mathcal{S}_f^{(p)}(h)$ are in fact $0_\varepsilon(e^{-S_0/h + \varepsilon/h})$ for each $\varepsilon > 0$ where $S_0 = \inf_{j \neq k} d(U_j,U_k)$

Witten [WI] mentions this fact (but only the $0(h^\infty)$ result) and a proof of this fact is also given in [MELR]. The result given in [HE-SJ]$_4$ is in fact better.

Proof.

Modulo $\tilde{O}(e^{-S_0/h})$ (in the sense of remark 4.2.2), we are reduced to the study of the Dirichlet problems attached to each well. Proposition (5.2.5) is a consequence of the:

Proposition 5.2.6.

Let $p \in [0,\dots,n]$ and $U_j$ a critical point of f. Let $\mathcal{S}^{(p)}_{f,M_j}(h)$ the Dirichlet realization of $\mathcal{S}_f^{(p)}(h)$ in $M_j$. Then :

(5.2.18) If $l_j \neq p$, $\mathrm{Sp}\ \mathcal{S}^{(p)}_{f,M_j}(h) \cap [0,Ah^{3/2}] = \phi$ for $h \in ]0,h_A]$

(5.2.19) If $l_j = p$, $\mathrm{Sp}\ \mathcal{S}^{(p)}_{f,M_j}(h) \cap [0,Ah^{3/2}] = \mu_j(h)$

and

(5.2.20) $\mu_j(h) = \tilde{\mathcal{O}}(e^{-2S_j/h})$ where $S_j = \min_{k \neq j} d(U_j,U_k)$

Proof

(5.2.18) and (5.2.19) are just a consequence of the general theory. So we are reduced to the proof of (5.2.20). Let us suppose that $l_j = p$ and let $\varphi_j$ a normalized eigenfunction associated to $\mu_j$ :

$$(5.2.21)\quad \begin{cases} (1) & \mathcal{S}_f^{(p)}(h)\,\varphi_j = \mu_j\,\varphi_j \\ (2) & \varphi_j \restriction_{\partial M_j} = 0 \end{cases}$$

Let us observe now that $d_f$ commute with $\mathcal{S}_f(h)$, in particular :

$$(5.2.22)\qquad d_f^{(p)}\,\mathcal{S}_f^{(p)}(h)\,\varphi_j = \mathcal{S}_f^{(p+1)}(h)\,(d_f^{(p)}\,\varphi_j) = \mu_j (d_f^{(p)}\,\varphi_j)$$

but, of course, there is no reason to hope that :

$$(d_f^{(p)}\,\varphi_j)\restriction_{\partial M_j} = 0$$

Let us cut by multiplying by the function $\chi_j$ introduced in (4.2.8). Then, using the decay of $(d_f^{(p)}\,\varphi_j)$, we get :

$$(5.2.23)\qquad (\mathcal{S}_f^{(p+1)}(h) - \mu_j(h))\,(\chi_j\,d_f^{(p)}\,\varphi_j) = \tilde{\mathcal{O}}(e^{-S_j/h})$$

Now, from (5.2.18), because $l_j(=p) \neq (p+1)$, we get :

$$\varepsilon_0\,h^{3/2}\,\|\chi_j\,d_f^{(p)}\,\varphi_j\| \leq d(\mu_j, \mathcal{S}_f^{(p+1)})\,\|\chi_j\,d_f^{(p)}\,\varphi_j\| = \tilde{\mathcal{O}}(e^{-S_j/h})$$

and finally :

$$(5.2.24)\qquad \|\chi_j\,d_f^{(p)}\,\varphi_j\| = \tilde{\mathcal{O}}(e^{-S_j/h}) = \|d_f^{(p)}\,\varphi_j\|$$

A similar argument gives that :

$$(5.2.25)\qquad \|d_f^{*(p-1)}\,\varphi_j\| = \tilde{\mathcal{O}}(e^{-S_j/h})$$

Let us now return to $(5.2.21)_{(1)}$ and take the scalar product with $\varphi_j$ ; then we get (see 5.2.8) :

$$(5.2.26)\qquad \|d_f^{(p)}\,\varphi_j\|^2 + \|d_f^{*(p-1)}\,\varphi_j\|^2 = \mu_j\,\|\varphi_j\|^2 = \mu_j$$

Using (5.2.24) and (5.2.25), this gives (5.2.20) and the end of the proof.

//

## §6 ON SCHRÖDINGER OPERATORS WITH PERIODIC ELECTRIC POTENTIALS.

### § 6-1 Generalities

We assume in this section that the potential V is $C^\infty$ on $R^n$ and periodic i.e :

$$(6.1.1) \quad V(x+\vec{a}_i) = V(x) \ , \quad x \in R^n$$

where $\vec{a}_i = a_i \cdot \vec{e}_i$ , $a_i > 0$ , $\vec{e}_i$ is the canonical basis in $R^n$.

The theory of the Schrödinger operator with periodic potentials is explained in section XIII-16 of Reed-Simon tome IV [RE-SI] (See also [WIL], [EA]). We just recall some basic facts.

We are interested to study the spectrum of

$$(6.1.2) \quad P(h) = -h^2 \Delta + V$$

particularly near the minimum of V.

We first note the existence of the translation operators $T_i$ $(i=1,\dots,n)$

$$(6.1.3) \quad (T_i u)\ (x) = u(x-\vec{a}_i)$$

which generate a discrete group G isomorph to $Z^n$ and which commute to P(h). This will permit us to decompose P(h) as an hilbertian integral over $[0,2\pi]^n = T^n$ :

$$(6.1.4) \quad P(h) = \int_{T^n}^{\oplus} P_\Theta(h)\, d\Theta$$

according to the irreducibles representations of G.

More explicitly, let us consider the operator $\mathcal{G}$ from $L^2(R^n)$ into $L^2(T^n;L^2(Q))$ defined on $\mathcal{S}(R^n)$ by :

$$(6.1.5) \quad (\mathcal{G}u)\ (\Theta,x) \overset{\text{def}}{\equiv} \sum_{\alpha\ Z^n} e^{-i\Theta\alpha}\ (T^\alpha u)\ (x)$$

$$\equiv (\mathcal{G}_\Theta\, u)\ (x)$$

where $T^\alpha = T_1^{\alpha_1} \dots T_n^{\alpha_n}$ , $Q = \prod_i \left[-\frac{a_i}{2}, \frac{a_i}{2}\right]$

One verifies easily that $\mathcal{G}$ is isometric, Computing $\mathcal{G}^*$, one verifies that $\mathcal{G}^*$ is injective on $L^2(T^n;\ L^2(Q))$. Then we have obtained that $\mathcal{G}$ is unitary from $L^2(R^n)$ into $L^2(\ T^n,\ L^2(Q))$. Using the naturaldecomposition of $L^2\ (\ T^n,\ L^2(Q)) = \int_\Theta^+ L^2(Q)$, the commutation of P and $T_j$, and going back through $\mathcal{G}^*$, this permits us to get the decomposition (6.1.4) corresponding to a decomposition of :

$$L^2(R^n) = \int_{T^n}^{\oplus} \mathcal{H}_\Theta\, d\Theta$$

where $\mathcal{H}_\Theta$ is identified to the $u \in L^2_{loc}(R^n)$ s.t

(6.1.6) $(T^\alpha u) = e^{i\Theta\alpha} u$

In this identification, the domain $D_\Theta$ of $P_\Theta(h)$ is given by

$u \in \mathcal{H}_\Theta$, s.t $P_\Theta u \in \mathcal{H}_\Theta$

and then given by :

(6.1.7) $D_\Theta = \{u \in H^2_{loc}, (T^\alpha u) = e^{i\Theta\alpha} u\}$

and for $u \in D_\Theta$:

(6.1.8) $P_\Theta(h) u = P(h) u$

The isometry between $L^2(Q)$ and $\mathcal{H}_\Theta$ is given by

$u \longrightarrow \tilde{u}_\Theta(x) = e^{-i\Theta\alpha} u(x+\alpha a)$ for $x \in Q-\alpha\vec{a}$

whose inverse is simply the restriction to Q.

In this correspondance, we can see the operator $P_\Theta$ with domain $D_\Theta$ as working on $L^2(Q)$ with domain :

(6.1.7)' $D_\Theta = \{u \in H^2(Q),\ u(x - \vec{a}_j) = e^{i\Theta_j} u(x)$

$$\frac{\partial u}{\partial x_j}(x - \vec{a}_j) = e^{i\Theta_j} \frac{\partial u}{\partial x_j}(x)$$

for the x s.t $x - \vec{a}_j$ and $x \in \partial Q$ a.e$\}$

(6.1.8)' $P_\Theta u = P u$ in $L^2(Q)$ for $u \in D_\Theta$

This can be seen with this formulation as a boundary problem in Q which can be seen as associated to the sesquilinear from

$$q_\Theta(u,v) = h^2 \sum_i \int \frac{\partial u}{\partial x_i} \cdot \overline{\frac{\partial v}{\partial x_i}} + \int_Q V\ u.\ \bar{v}\, dx$$

for $u \in V_\Theta$ where

$V_\Theta = \{u \in H^1(Q),\ u(x - \vec{a}_j) = e^{i\Theta} u(x)$

for almost all x s.t $x-\vec{a}_j$ and $x \in \partial Q\}$

To see that for each $\Theta$, $P_\Theta$ is with compact resolvent, the best is probably to make the change of function :

(6.1.9) $u \longrightarrow e^{-i\Theta x} v$

which sends $D_\Theta$ onto $D_0$ and $P_0$ onto

(6.1.10) $\check{P}_\Theta = \sum_j (hD_{x_j} - h\Theta_j)^2 + V$

$\check{P}_\Theta$ can then be seen as an operator on $L^2(T_a{}^n)$ in the identification of Q with the the torus. The domain is then $H^2(T_a{}^n)$ whose injection is compact in $L^2(\Pi_a^n)$.
Of course, $\check{P}_\Theta$ (h) and $P_\Theta$ are unitary equivalent.

$P_\Theta$ (or $\check{P}_\Theta$ ) has a discrete spectrum composed from a sequence of eigenvalues with finite multiplicity

$$\lambda_0^\Theta(h) \leq \lambda_1^\Theta(h) \leq \ldots\ldots \leq \lambda_k^\Theta(h) \leq \quad \ldots$$

tending to $\infty$ .
We denote by $(\phi_k^\Theta)$ a corresponding orthonormal basis of eigenvectors.
By the minimax principle, the eigenvalues depend continuously of $\Theta$ and the spectrum of P(h) is a union of bands $B_k(h)$ :

(6.1.11) $$S_p \quad P(h) = \bigcup_{k \in N} B_k(h)$$

(6.1.22) $$B_k(h) = \bigcup_{\Theta \in T^n} \lambda_k^\Theta(h)$$

Of course, these bands can overlap. In this section, we will concentrate our study to the study of $B_0(h)$ and prove under suitable hypothesis that $B_0(h)$ does'nt overlap with the other bands and that it is possible to give for h small enough the asymptotic behavior of the length of $B_0(h)$.
This will be the purpose of § 6-2.

## § 6-2 Semi-classical study of the first eigenvalue of $P_\Theta$ .

According to § 6-1 , we are reduced to the study of $P_\Theta$

where

(6.2.1) $$\begin{cases} D_\Theta = \{u \in H^2_{loc}(R^n) \ , \ T^\alpha u = e^{i\Theta\alpha} u \} \\ P_\Theta u = (-h^2\Delta + V)u \quad \text{in } R^n \end{cases}$$

We shall assume that :

(6.2.2) $$\begin{cases} \text{V admits a non degenerate minima at the points of the lattice} \\ L = \{ (\sum_i \alpha_i \vec{a}_i) \ (a \in Z^n) \text{ and there are no other minima.} \end{cases}$$

If one thinks to the equivalent problems on $L^2(Q)$ (see (6.1.7)' , (6.1.8)' ,

this is essentially a one well problem which is relevant of the technics developped in § 2 to 4. We shall follow the approach of Outassourt [OU] (but see also $[HA]_2$ , $[SI]_6$, [CAR] for connected approachs or results)
To normalize the problem we assume that :

(6.2.3) $V(0) = 0$

The first step is as in section 3.1 to have an a priori estimate of the decay of the eigenfunctions $\phi_k^\Theta(h)$ corresponding to eigenvalues $\lambda_k^\Theta(h) \in I(h)$ where I(h) is an intervall inside [0,Ch] for some fixed constant C.

Proposition 6.2.1

Let u a normalized eigenfunction of $P_\theta$ (h) associated to $\lambda^0(h) \in I(h)$. Then for all $\varepsilon > 0$, we have :

$$(6.2.4) \quad \| \nabla [e^{\phi/h} u] \|^2_{L^2(\Omega)} + \| e^{\phi/h} u \|^2_{L^2(\Omega)} = O_\varepsilon (e^{\varepsilon/h})$$

where $\phi(x) = \inf_{\alpha . \vec{a} \in L} d_V (\alpha a, x)$ for $x \in \Omega$

($d_V$ is the Agmon distance associated to V)

Moreover $O_\varepsilon$ can be chosen independently of $\theta \in T^n$.

Proof The proof is quite similar to the proof of proposition 3.3.1. We just mention the significative new steps :

Using (6.1.9) and (6.1.10), we prefer to prove the same property for an eigenfunction v of the operator $\check{P}_\theta$:

$$(6.2.5) \quad \check{P}_\theta (h) = \sum_j (hD_{x_j} - h\,\theta_j)^2 + V(x)$$

which we consider as an unbounded operator on $L^2(T^n_a)$ ($T^n_a$ is a compact manifold). We have just to prove the corresponding inequality to 6.2.4 for v with $\phi$ replaced by $\check{\phi}$ where $\check{\phi}$ is the Agmon distance associated to V considered as a $C^\infty$ function on $T^n_\alpha$. The only difference with the case studied in §3 is the presence of $(h\,D_{x_j} - h\theta_j)$ instead of $(hD_{x_j})$. It is then clear, that we need to prove the analogue of theorem 3.1.1 :

Proposition 6.2.2

Let $\rho$ be a real Lipschitzian function on $T^n_a$, then for each $v \in C^2(T^n)$ we have :

$$(6.2.6) \quad \mathrm{Re} \int_{T^n_a} e^{2\rho(x)/h} [\check{P}_\theta(h) - \lambda_\theta(h) v(x)]\, \bar{v}(x)\, dx$$
$$= \|(h\nabla - ih\theta)\, e^{\rho/h}\, v\|^2 + \int e^{2\rho/h} (V - |\nabla\rho|^2 - \lambda_\theta(h))\; v.\bar{v}\, dx$$

Then (6.2.4) is deduced from (6.2.6) along the same arguments as in § 3.

The second step is to introduce a one well reference problem in $L^2(R^n)$. Let $W_\eta$ be a positive $C^\infty$ function on $R^n$ with compact support in the Ball $B(0,\eta/2)$ for some $\eta > 0$ (the ball is taken with respect to the Agmon distance and $\eta$ will be chosen later arbitrarily small) and equal to 1 on $B(0,\eta/4)$.

Let us consider the operator :

$$(6.2.7) \quad \tilde{P}_\eta (h) = P(h) + \sum_{\substack{\alpha \neq 0 \\ \alpha \in Z^n}} W_\eta \,(x - \sum_i \alpha_i \vec{a}_i)$$

$\tilde{P}_\eta$(h) has one well at 0 and coincides with P(h) on $B(0, S_0 - \eta/2)$ where :

(6.2.8) $$S_o = \min_{\alpha \in \mathbb{Z}^n \setminus 0} d_V(0, \sum_i \alpha_i \vec{a}_i)$$

It is not difficult to see that the minimum is obtained for a finite number of $\alpha$.

Let $\tilde{\lambda}(h)$ be the first eigenvalue of $\tilde{P}(h)$. According to § 2-4, $\tilde{\phi}(h)(x)$ (the corresponding first eigenfunction) is "well known", decreases in $B(0, S_o - \eta/2)$ like $O(h^{-N}) \; e^{-d_V(x,0)/h}$, has a B.K.W approximation in "good domains" and if we take $I(h) = [\mu_o h - \tilde{c}h^2 , \mu_o h + \tilde{c}h^2]$ with $\mu_o h$ the first eigenvalue of the harmonic approximation at the Bottom, $I(h)$ contains only $\tilde{\lambda}(h)$ and is at distance $\frac{1}{c}$ h of the other eigenvalues of $\tilde{P}(h)$. Let $\chi_\eta$ a $C^\infty$ function with support in $B(0, S_o - \frac{3\eta}{4})$ and equal to 1 on $B(0, S_o - \eta)$ and define :

(6.2.4) $$\tilde{\psi} = \chi \tilde{\phi}$$

Then we have :

(6.2.10) $$P(h)\tilde{\psi} = \tilde{\lambda}(h)\tilde{\psi} + \tilde{r}$$

$$\text{with } \tilde{r} = [P, \chi]\tilde{\phi} \qquad (= O(e^{-\frac{S_o}{h} + \frac{\eta}{h}})$$

Let $F_\Theta(h)$ be the eigenspace of $P_\Theta(h)$ associated to the interval $I(h)$. If $\phi_\Theta$ is an eigenvector in $F_\Theta$ associated to $\lambda_\Theta$ in $I(h)$, then if $\tilde{\chi}$ is a $C_o^\infty$ function equal to 1 in $B(0, \eta/2)$ with support in $B(0, \eta)$

$$\tilde{P}(\tilde{\chi}\phi_\Theta) = \lambda_\Theta(\tilde{\chi}\phi_\Theta) + r_\Theta \qquad \text{with } r_\Theta = O(e^{-\varepsilon_o h}) \text{ in } L^2$$

From this, we deduce by Proposition 4.1.1 , that $\dim F_\Theta < 1$ and that $\lambda_\Theta$ is necessarily exponentially near $\tilde{\lambda}(h)$. Changing eventually a little $I(h)$, we get that Card $(S_p P_\Theta \cap I(h)) \leq 1$ and that $\partial I(h)$ is at distance $\geq \varepsilon_1 h^2$ of $S_p P_\Theta(h)$ where $\varepsilon_1$ is can be chosen independant of $\Theta$.

Let us now prove the converse : $\dim F_\Theta(h) > 1$ by constructing a quasi-mode for $P_\Theta(h)$ with the help of $\tilde{\psi}$.

Let us consider :

(6.2.11) $$\tilde{\psi}_\Theta(x) = \sum_{\alpha \in Z^n} e^{-i\Theta\alpha} (T^\alpha \tilde{\psi})(x) \qquad \text{for } x \in R^n$$

The sum in the definition of $\tilde{\psi}_\Theta$ is finite and $\tilde{\psi}_\Theta \in \mathcal{H}_\Theta$. Moreover :

(6.2.12) $$P_\Theta \tilde{\psi}_\Theta(x) = \tilde{\lambda}(h) \tilde{\psi}_\Theta + \sum_{\alpha \in Z^n} e^{-i\Theta\alpha} (T^\alpha \tilde{r})(x)$$

$P_\Theta \tilde{\psi}_\Theta(x) = \hat{\lambda}(h)\tilde{\psi}_\Theta + \tilde{r}_\Theta$

It is then clear that $\tilde{r}_\Theta = \mathcal{O}(e^{-\frac{S_o}{h}+\frac{\eta}{h}})$ uniformly with respect to $\Theta$.

From this we get from Proposition 4.1.1 , that there exists a unique $\lambda^\Theta(h)$ (which is $\lambda_o^\Theta(h)$) s.t

(6.2.13) $\lambda_o^\Theta(h) - \tilde{\lambda}(h) = \mathcal{O}(e^{-\frac{S_o}{h}+\frac{\eta}{h}})$

$$d(\mathbb{C}\tilde{\psi}_\Theta, F_\Theta) = \mathcal{O}(e^{-\frac{S_o}{h}+\frac{\eta}{h}})$$

Then following the arguments of § 4-3, we get

<u>Lemma 6.2.3</u> Under the preceeding hypotheses.

(6.2.14) $\lambda_o^\Theta(h) = \tilde{\lambda}(h) + (\tilde{r}_\Theta/\tilde{\psi}_\Theta) + O(e^{-\frac{2S_o}{h}+2\eta/h})$

It remains to compute $(\tilde{r}_\Theta/\tilde{\psi}_\Theta)$ and to relate it to known quantities related $\tilde{r}$ and $\tilde{\psi}$.

According to the decay properties of $\tilde{\phi}$, $\tilde{r}$ and because the support of $\tilde{r}$ is contained in $B(0, S_o - \frac{\eta}{2})$ $B(0, S_o-\eta)$, we get according to (6.2.11) and (6.2.12) :

(6.2.15) $(\tilde{r}_\Theta/\tilde{\psi}_\Theta)_{L^2(Q)} = \sum_{\delta\in Z^n\setminus 0} e^{-i\delta.\Theta}(T^\delta\tilde{r}/\tilde{\psi})_{L^2(R^n)} + O(e^{-\frac{2S_o}{h}+\frac{2\eta}{h}})$

Observing that $\tilde{\psi}$, $\tilde{r}$ are real and that $(T^\delta\tilde{r}/\tilde{\psi}) = (T^{-\delta}\tilde{r}/\tilde{\psi})$, we get

(6.2.16) $(\tilde{r}_\Theta/\tilde{\psi}_\Theta)_{L^2(Q)} = \sum_{\delta\in Z^n\setminus 0} \cos\delta\Theta \; (T^\delta\tilde{r}/\tilde{\psi})_{L^2(R^n)} + O(e^{-\frac{2S_o}{h}+2\eta/h})$

and finally the :

<u>Theorem 6.2.4</u> Under the hypothesis (6.2.1), $P_\Theta(h)$ admits in the intervall $I(h) = [\mu_o h - Ch^2 u_o h + Ch^2]$ a unique eigenvalue $\lambda_o^\Theta(h)$ which is simple (and the smallest) and verifies :

(6.2.17) $\lambda_o^\Theta(h) = \tilde{\lambda}(h) + \sum_{\delta\in Z^n\setminus 0} \cos(\Theta\delta)\; (T^\delta\tilde{r}/\tilde{\psi})_{L^2(R^n)} + O(e^{-\frac{2S_o}{h}+\frac{2\eta}{h}})$

where $\tilde{\lambda}(h)$ is the first eigenvalue of the reference problem $\tilde{P}(h)$.

Moreover $O(e^{-\frac{2S_o}{h}+\frac{2\eta}{h}})$ is uniform with respect to $\Theta$.

This theorem has been obtained by A. Outassourt (Th 4.5 in [OU])

Remark 6.2.5

This result seems to depend of the reference problem in particular of $\eta$ . But according to the considerations of §3 and 4 we know that for two different choices of $\eta:\eta',\eta$

$$\tilde{\lambda}_{\eta'}(h) - \tilde{\lambda}_{\eta''}(h) = O\left(e^{-\frac{2S_0}{h} + 2\frac{\sup(\eta,\eta')}{h}}\right)$$

The measure of $B_0(h) = \bigcup_{\theta\in T^n} \lambda_0^\theta(h)$ is then reduced to the computation of $(T^{\delta}\tilde{r}/\tilde{\psi})$

To compute the main contribution, we define :

$$\Lambda = \{\delta \in Z^n \setminus 0, \text{s.t} \quad d_V(0, \sum \delta_i \vec{a}_i) = S_0$$

M is a finite set invariant by $\delta \longrightarrow -\delta$

Let us define $S'_0 = \inf_{\delta\in(Z^n\setminus 0)\setminus\Lambda} d_V(0,\sum \delta_i \vec{a}_i)$

Then we write (6.2.17) on the form :

$$\text{(6.2.18)} \quad \lambda_0^\Theta(h) = \tilde{\lambda}(h) + \sum_{\delta\in\Lambda} \cos(\Theta\delta)\,(T^{\delta}\tilde{r}|\tilde{\psi})_{L^2(R^n)} + \tilde{O}(\exp(-S'_0/h))$$

Let us consider more carefully :

$$(T^{\delta}\tilde{r}|\tilde{\psi}) = (T^{-\delta}\tilde{r}/\tilde{\psi}) = (\tilde{r}/T^{\delta}\tilde{\psi})$$

$(T^{\delta}\tilde{\psi})$ is localized in the well $(\sum \delta_i \vec{a}_i)$ and we are <u>exactly</u> in the same situation as in the computation of $W_{\alpha,\beta}$ in section 4.3 where the computation was given in terms of $\phi_\alpha$ and $\phi_\beta$ (here $\tilde{\phi}$ and $\tilde{\phi}_\delta = T^\delta \tilde{\phi}$). We then get (see theorem 4.3.4)

$$\text{(6.2.19)} \quad (\tilde{r}\,|T^{\delta}\tilde{\psi}) = h^2 \int_{\Gamma_\delta} (\tilde{\phi} \cdot \frac{\partial \tilde{\phi}_\delta}{\partial n} - \tilde{\phi}_\delta \cdot \frac{\partial\tilde{\phi}}{\partial n})\, dS_{\Gamma_\delta} \quad (\text{modulo } O(h^\infty)e^{-S_0/h})$$

where $\Gamma_\delta$ is defined as after 4.3.24 and is between the well

$U_0 = \{0\}$ and $U_\delta = \{\sum_i \delta_i \vec{a}_i\}$

Using the theorem 4.4.6, we then get :

$$\text{(6.2.20)} \quad \frac{1}{C} h^{1/2} \leq e^{S_0/h} \,|(\tilde{r}/T^{\delta}\tilde{\psi})| \leq Ch^{1-n/2}$$

for some constant C>0.

Moreover under suitable hypotheses on minimal geodesics between $U_0$ and $U_\delta$ (i.e Hypothesis 4.4.42) (see theorem 4.4.8) , we can get :

$$\text{(6.2.21)} \quad e^{S_0/h}(\tilde{r}/T^{\delta}\tilde{\psi}) = h^{1/2}\,(\sum_{m\in N} b_\delta^m h^m)\, e^{-S_0/h} \quad (\text{modulo } O(h^\infty)\, e^{-S_0/h})$$

From formula (6.2.18) and (6.2.20) , if we remark that all the contributions $(T^{\delta}\tilde{r}/\tilde{\psi})$ have for h small enough the same sign we get the :

Theorem 6.2.6 (Cf Outassourt [OU])

Under the hypothesis (6.2.1) , the first band of P(h) : $B_o(h)$ is contained in I(h) (and this the only one) and :

$$\text{(6.2.22)} \quad \text{length } (B_o(h)) = 2 \sum_{\delta \in \Lambda} \left| (T^{\delta}\tilde{r}/\tilde{\psi}) \right| + \mathcal{O}(e^{-S_o/h})$$

$$\text{(6.2.23)} \quad \frac{1}{C_1} \leq h^{1/2} \text{ length } (B_o(h)) \; e^{S_o/h} \leq C_1 \; h^{1-n/2}$$

and if (4.4.42) is satisfied for all geodesics between $U_o$ and $U_{\delta}$ for $\delta \in \Lambda$ , we get :

$$\text{(6.2.24)} \quad \text{length } (B_o(h)) \equiv h^{1/2} \Big( \sum_{m \in \mathbb{N}} b_m \; h^m \Big) \; e^{-S_o/h} \quad (\text{modulo } O(h^{\infty}) e^{-S_o/h})$$

with $b_o \neq 0$

We deduced from (6.2.23) the following corollary obtained by B.Simon in $[SI]_6$

Corollary 6.2.7

$$\lim_{h \to 0} -h \text{ Log } (\text{length } (B_o(h))) = S_o$$

Remark 6.2.8

The case n = 1 was studied by Harrell $[HA]_2$ . In this case, we can get a explicit result in the spirit of § 4.5.

Remark 6.2.9 In Carlson [CAR] is developped another method which corresponds to a general reduction for problems with an infinite number of wells. This method has the advantage to work also in non periodic cases.

## §7 ON SCHRÖDINGER OPERATORS WITH MAGNETIC FIELDS.

### § 7.0 Introduction

In this section we want to present more recent results on the spectrum of the Schrödinger operator with magnetic fields. Although the theory is quite interesting on manifolds , we will only study the case of $R^n$ or of an open set in $R^n$. We refer to the basic references [A.H.S] and [HU] for a presentation or a survey of many interesting problems, see also [C.F.K.S]

The starting point is the 1-form :

(7.0.1) $\quad \omega_A = \sum_j A_j \, dx_j$

(or the corresponding vector $\vec{A} = (A_1, \dots, A_n)$ , called the magnetic potential vector) where $A_j$ is in $C^\infty(R^n)$

and the electric potential

7.0.2 $\quad V \in C^\infty(R^n)$ which is semi-bounded : $V > -C_0$

The magnetic field is by definition the two-form :

7.0.3 $\quad \sigma_B = d\omega_A = \sum_{\substack{j,k \\ j<k}} B_{jk} \, dx_j \wedge dx_k$

where $B_{jk}$ can be identify to a corresponding antisymmetric matrix. The Schrödinger operator with magnetic fields in then defined by

7.0.4 $\quad P_A(h) = \sum_{1 \leq j \leq n} (hD_{x_j} - A_j)^2 + V$

Let us first remark the following important property called gauge invariance :

Lemme 7.0 If $\vec{A}'$ and $\vec{A}''$ are two magnetic potential vectors such that $d\omega_{A'} = d\omega_{A''}$ , then there exists a unitary operator U of the form :

$L^2 \ni f \longrightarrow (Uf) = e^{i\phi/h} f \qquad$ (for some $\phi \in C^\infty(R^n)$)

s.t

7.0.5 $\quad P_{A'}(h)\, U = U\, P_{A''}(h) \quad$ on $C_0^\infty(R^n)$

This is just the consequence of the fact that on $R^n$ a closed 1-form is exact so :

$\omega_{A'} - \omega_{A''} = d\phi$

(of course , it is no larger true on a manifold (like $S^1$) or on a open set (like $R^2 \setminus B(0;\delta)$) where interesting phenomena appear as we shall see in § 7.2 and 7.3)
This gauge invariance explains why (as in the classical theory of electromagnetism) we will frequently give hypotheses on B (and not on A) , when we are interested with problems invariant by unitary transformation like in spectral theory.

Let us recall also that, under the hypotheses 7.0.1 and 7.0.3, $P_A(h)$ is essentially self adjoint on $L^2(R^n)$ starting from $C_o^\infty(R^n)$. The domain of his self adjoint extension is given by :

(7.0.6) $D(P_A(h)) = \{u \in L^2(R^n)\ ,\ (h D_{x_j} - A_j)u \in L^2(R^n),\ P_A(h)u \in L^2(R^n),$
$(V+c_o)^{1/2} u \in L^2(R^n)\}$ and $C_o^\infty(R^n)$ is dense in $D(P_A(h))$.

In the case of open set $\Omega$ with regular bounded boundary $\partial\Omega$. We start with the sesquilinear form :

$$q_A(u,v) = \sum_j \int (hD_{x_j} - A_j)u.\ \overline{(hD_{x_j} - A_j)v}\ dx + \int V\ u.\bar{v}\ dx \text{ defined on}$$

$$H^1_{V,A}(\Omega) = \overline{C_o^\infty(\Omega)}^{\|\ \|_{H^1_{V,A}}}$$

where $\|u\|^2_{H^1_{V,A}(\Omega)} = \|u\|^2_{L^2(\Omega)} + \sum_j \|(hD_{x_j} - A_j)\ u\|^2_{L^2(\Omega)} + \|(V+c_o)^{1/2}\ u\|^2_{L^2(\Omega)}$

and the domain of the associated operator $P_A^\Omega(h)$ on $L^2(\Omega)$ is given by

$$D(P_A^\Omega(h)) = \{u \in H^1_{O,A}(\Omega),\ (V+c_o)^{1/2} u \in L^2(\Omega),\ P_A(h)\ u \in L^2(\Omega)\}$$

(7.0.7)

$$P_A^\Omega(h)\ u = P_A(h)u \quad \text{in } \mathcal{D}'(\Omega)$$

A core for $P_A^\Omega(h)$ is given by the u in $C_o^\infty(\bar{\Omega})$ s.t. $u \wedge_{\partial\Omega} = 0$

We will consider in this section three problems.

The first problem is to know if the resolvent of $P_A^\Omega(h)$ is compact. It is classical if V tends to $\infty$ as $|x| \longrightarrow \infty$. As mentionned in § 1, it can occur that the resolvent is compact for some B, in the case V = 0. We will present in § 7-1 a result of Helffer-Mohamed [HE-MO] extending previous results by [A-H-S] and other authors giving a general criterion of compactness of the resolvent.

The second problem is to analyze the effect of a magnetic field on the first eigenvalue. In the case when $\Omega$ is non simply-connected, we shall see effects where not only $\sigma_B$ plays a role but also $\omega_A$ through the integral of $\omega_A$ along closed path. This is connected with the Aharonov-Bohm effect discovered in 1959 [AH-BO].

The third problem is related to the study of the multiplicity of the first eigenvalue. It is well known (see [RE-SI]) that the first eigenvalue is of multiplicity one. One presents here two examples where the multiplicity is two (see [LA-O'CA] and [A.H.S] for other examples).

This problem is far to be completely understood in general.

## § 7.1 A criterion of compact resolvent

In this section h is fixed and we will take it equal to 1 and $\Omega = \mathbb{R}^n$.
If one wants to prove the compactness of the resolvent, it is sufficient to prove (cf 7.0.6) that :

(7.1.1) $D(P_A) \hookrightarrow L^2_\rho$ where $L^2_\rho = \{u \text{ mesurable s.t } \int |u|^2 \rho(x)^2 d < \infty\}$

for some real continuous function $\rho$ tending to $\infty$ when $|x| \to \infty$

Indeed, it is clear that :

(7.1.2) $D(P_A) \hookrightarrow H^2_{loc}$

and by usual criteria of compactness, the injection of $D(P_A)$ in $L^2$ is compact. According to the density of $C_0^\infty(R^n)$ in $D(P_A)$, we are then reduced to find simple criteria to get some function $\rho(x)$ tending to $\infty$ when $|x| \to \infty$ and a constant $C>0$ s.t for each $u \in C_0^\infty(R^n)$ we have :

(7.1.3) $\int |u(x)|^2 \rho(x)^2 dx < C \left(\|P_A u\|^2_{L^2} + \|u\|^2_{L^2}\right.$

Different criteria were known (cf [A.H.S] , $[IWA]_2$ , [HE-MO]). We shall present here the criterion given in [HE-MO]. To understand the meaning of this criterion, let us first treat a simple example. Let us consider on $R^2$ the case when :

$A_1(x) = -x_1^2 \cdot x_2 \quad ; A_2(x) = x_1 x_2^2 \quad ; V = 0.$

We first observe that :

$$[D_{x_1} - A_1 , D_{x_2} - A_2] = i(\partial_{x_1} A_2 - \partial_{x_2} A_1) = i\,(x_1^2 + x_2^2)$$

and we get :

$$\|(x_1^2 + x_2^2)^{1/2} u\|^2_{L^2} = \int (x_1^2 + x_2^2)\,|u|^2\, dx = i^{-1} \int ([D_{x_1} - A_1, D_{x_2} - A_2] u(x) . \overline{u(x)}\, dx$$

$$= i^{-1} \int (D_{x_2} - A_2) u . (D_{x_1} - A_1) u . dx - i^{-1} \int (D_{x_1} - A_1)\, u\, \overline{(D_{x_2} - A_2)\, u}\, d\,x$$

$$\leqslant \|(D_{x_1} - A_1)\, u\|^2 + \|(D_{x_2} - A_2)\, u\|^2$$

$$\leqslant C\, (\|P_A u\|^2 + \|u\|^2)$$

We then obtain immediatly (7.1.3) with $\rho(x) = (x_1^2 + x_2^2)^{1/2}$.

One could think that the crucial property is the fact that $B(x) \to \infty$ . It is not exactly true. Indeed let us consider the second example where :

$A_1 = x_2\, x_1^2 \quad , A_2 = x_1\, x_2^2$

Then we get : $B(x_1, x_2) = x_2^2 - x_1^2$ and B doesn't tend to infinity.

But as a consequence of the results of [HE-NO] , we have compact resolvent. The crucial fact is here that :

$$|[D_{x_1} - A_1, [D_{x_1} - A_1 , D_{x_2} - A_2]]| + |[D_{x_2} - A_2 , [D_{x_1} - A_1 , D_{x_2} - A_2]]| \text{ tends to } \infty \text{ as}$$

$x$ tends to $\infty$ . To generalize this example, let us introduce the following notations Let us assume that:

(7.1.4) $$V(x) = \sum_{1 \leqslant j \leqslant p} V_j(x)^2$$ (with $V_j$ real, $C^\infty$ ) and for $l \in N$, let us

introduce :

(7.1.5) $$m_l(x) = \sum_{|\alpha|=l} |\partial_x^\alpha V_j| + \sum_{\substack{|\alpha|=l-1 \\ i,j}} |\partial_x^\alpha B_{ij}|$$

(For $l = 0$, forget the second term of the right hand side) and for $r \in N$

(7.1.6) $$m^r(x) = 1 + \sum_{l=0}^{r} m_l(x) .$$

The following theorem is a particular case of a more general theorem in [HE-MO]:

Theorem 7.1.1

Let us assume (7.0.1) , (7.1.4) and that for some $r \in N$, some $C>0$ :

(7.1.7) $$m_{r+1}(x) \leqslant C \; m^r(x)$$

(7.1.8) $$m^r(x) \xrightarrow[|x| \to \infty]{} \infty$$

then $P_A(h)$ has compact resolvent.

Remark 7.1.2

The case $r = 0$ is a particular case of the classical result that if $V(x) \to \infty$ then $P_A(h)$ has compact resolvent for each A satisfying 7.0.1

Proof of theorem 7.1.1

Let us first recall the existence of a partition of unity on $R^n$ such that

(7.1.9)
$$\begin{cases} \sum_k \chi_k(x)^2 = 1 \\ \sum |\partial_x^\alpha \chi_k(x)|^2 \leqslant C_\alpha \\ \text{Each point } x \text{ is in at most } M \text{ supports of } \chi_k \\ \text{Supp } \chi_k \text{ is compact, of diameter less than 1.} \end{cases}$$

We denote by $y_k$ a point s.t $\chi_k(y_k) \neq 0$

Let $\tau$ be a parameter precized later s.t $0<\tau<1$. And let us introduce :

$$(7.1.10) \quad \chi_k^\tau(x) = \chi_k(x/\tau)$$

Then $\operatorname{supp} \chi_k^\tau \subset B(\frac{y_k}{\tau}, \tau)$

Let us now define :

$$(7.1.11) \quad \psi(x) = \sum_k m^r(\frac{y_k}{\tau})\,(\chi_k^\tau)^2(x)$$

According to (7.1.7), there exists a constant C such that for x $\operatorname{Supp} \chi_k^\tau$, $\tau < 1$

$$| m^r(\frac{y_k}{\tau}) - m^r(x) | < C\tau \sup_{y \in B(y_k/\tau, \tau)} m^r(y)$$

For $\tau$ small enough (independant of k!) we then get the existence of $C_o > 0$ and $C_1 > 0$ s.t :

$$C_o\, m^r(x) \leqslant m^r(\frac{y_k}{\tau}) \leqslant C_1\, m^r(x)$$

for x in the ball $B(\frac{y_k}{\tau}, \tau)$

We then deduce :

$$(7.1.12) \begin{cases} C_o\, m^r(x) \leqslant \psi(x) \leqslant C_1\, m^r(x) \\ \forall x \in R^n \end{cases}$$

We fix now $\tau$ s.t (7.1.12) is satisfied, and we have now the existence of constants $C_\alpha$ $(\alpha \in N^n)$ s.t :

$$(7.1.13) \begin{cases} |D_x^\alpha \psi(x)| < C_\alpha\, \psi(x) \\ \forall x \in R^n \end{cases}$$

The proof of theorem 7.1.1 was inspired by technics of Kohn [KO] introduced to give an alternative proof of the hypoellipticity of the Hörmander's operators : $\sum X_j^2$

For all s>0, let us introduce :

$$(7.1.14) \begin{cases} M^s \text{ is the space of the } C^\infty \text{ functions } T \text{ s.t} \\ \begin{cases} \exists C_s > 0 \text{ s.t } \forall u \in C_o^\infty(R^n) \\ \|\psi^{-1+s}\, T.\, u\| < C_s\, ((P_A u/u) + \|u\|^2) \end{cases} \end{cases}$$

Let us also introduce :

$$(7.1.15)\begin{cases} L_j = D_j - A_j & \text{if } j = 1,\dots,n \\ L_j = V_{j-n} & \text{if } n<j\leq n+p \end{cases}$$

and recall the basic estimate :

$$(7.1.16)\quad \sum_j \|L_j u\|^2 \leq (P_A u/u) \qquad \forall u \in C_o^\infty(R^n)$$

We shall now prove :

(7.1.17) $V_j \in M^1$ (direct consequence of 7.1.16)

(7.1.18) $[L_k , L_j] \in M^{1/2}$

and

(7.1.19) If $T \in M^S$ and $|\partial_x^\alpha T| < C_\alpha \psi$ for $|\alpha|=1$ or 2, then $[L_k,T] \in M^{S/2}$

Then it is clear that $\psi(x) \in M^{2^{-r}}$. Indeed we deduce from (7.1.18) and (7.1.19) that : $\partial_x^\alpha B_{jk} \in M^{2^{-(|\alpha|+1)}}$ and from (7.1.17) and (7.1.18) that : $\partial_x^\alpha V_j \in M^{2^{-|\alpha|}}$

It remains then to prove (7.1.18) and (7.1.19)

<u>Proof of 7.1.18</u>

For all u in $C_o^\infty(R^n)$ and each j,k, we write :

$$\|\psi^{-\frac{1}{2}} [L_k ; L_j] u\|^2 = \langle L_j u / \psi^{-1}[L_k,L_j] L_k u\rangle - \langle L_k u / \psi^{-1}[L_k,L_j]u\rangle$$

$$+\langle L_j u / [L_k ; \psi^{-1}[L_k,L_j]]u\rangle$$

$$- \langle L_k u / [L_j ;\psi^{-1} [L_k,L_j]] u\rangle$$

If we observe that $\psi^{-1}[L_k,L_j]$ and $[L_k ; \psi^{-1} [L_j,L_k]]$ are bounded, according to (7.1.12) and (7.1.13), we obtain :

$$(7.1.20)\quad \|\psi^{-1/2} [L_K ; L_j] u\|^2 \leq C(\|L_k u\|^2 + \|L_j u\|^2 + \|u\|^2)$$

and finally (7.1.18) according to (7.1.16).

<u>Proof of 7.1.19</u>

Let $T \in M_S$. For each k, we can write :

$$\| \psi^{-1+s/2} [L_k;T] u \|^2 = \langle \psi^{-1+s} Tu / \psi^{-1} [L_k,T] L_k u\rangle$$

$$- \langle L_k u / \psi^{-1} [L_k;T] \psi^{-1+s} Tu\rangle$$

$$+ \langle \psi^{-1+s} T u / \psi^{1-s} [L_k; \psi^{-2+s} [L_k;T]] u \rangle$$

If one observes that $\psi^{1-s} [L_k ; \psi^{-2+s} [L_k;T]]$ and $\psi^{-1} [L_k,T]$ are bounded , we get:

$\|\psi^{-1+s/2} [L_k;T] u\|^2 \leq C \left( \|\psi^{-1+s} Tu\|^2 + \|L_k u\|^2 + \|u\|^2 \right)$ and consequently 7.1.19.

<u>Remark 7.1.3</u>

Let us recall that connected interesting were obtained by Avron-Herbst-Simon [A.H.S], Colin de Verdière $[CDV]_2$ , Dufresnoy [DU], A.Mohamed [MO], Iwatsuka $[IW]_{1,2}$ B.Simon $[SI]_3$ , etc ...

## § 7.2 Spectral effects due to the flux of the magnetic fields

### 7.2.1 A qualitative result (h fixed)

As we have seen in section 7.0 (lemme 7.0.1), the gauge invariance suggests that only the magnetic field is relevant in spectral theory and not the potential vector $\vec{A}$. This is true in $R^n$ but it is not longer true in non simply connected open sets in $R^n$ (or more generally on non simply connected manifolds). The following theorem was conjectured and (almost) proved by R.Lavine and O'Caroll [LA-O'CA]

**Theorem 7.2.1.1**

Let $\Omega$ be a connected regular open set with bounded (eventually empty) boundary in $R^n$. Let us assume (to simplify) that $P_A^\Omega(h)$ and $P_O^\Omega(h)$ are with compact resolvent.

Let us denote by $\lambda_O^\Omega(h)$ (resp. $\lambda_A^\Omega(h)$) the first eigenvalue of the Dirichlet realization of $P_O(h)$ in $\Omega$ (respectively $P_A(h)$) $P_O^\Omega(h)$ (resp. $P_A^\Omega(h)$).

Then the following properties are equivalent :

(i) $\lambda_A^\Omega(h) = \lambda_O^\Omega(h)$

(ii) $P_A^\Omega(h)$ and $P_O^\Omega(h)$ are unitary equivalent

(iii) ⓐ $B=0$ in $\Omega$

ⓑ For all closed path $\gamma$ in , $\frac{1}{2\Pi h}\int_\gamma \omega_A \in Z$

**Proof of theorem 7.2.1.1**

(ii) $\Longrightarrow$ (i) is trivial

(iii) $\Longrightarrow$ (ii)

Let $\phi^R$ a solution in $\Omega^R$ (the universal simply-connected covering of $\Omega$ ) of $d\phi^R = \omega_A^R$ which exists from $(iii)_a$.

Condition $(iii)_b$ implies that $e^{i\frac{\phi^R(x)}{h}}$ a priori defined on $\Omega^R$ depends only on the projection on $\Omega$. The natural function, we get on $\Omega$ , defines by multiplication the unitary equivalence.

The proof of (i) $\Longrightarrow$ (iii) depends of an identity due to Lavine - O'Caroll [LA-O'CA]

Let $u_O$ the normalized (in $L^2(\Omega)$) positive first eigenfunction of $P_O^\Omega(h)$. It is well known (see [RE-SI]) that $u_O \in C^\infty(\Omega)$ is strictly positive in $\Omega$ and that the corresponding first eigenvalue is simple. This identity is easily obtained by integration by parts :

$\forall \phi \in C_o^\infty(\Omega)$ :

$$(7.2.1.1) \quad \left\| (h\nabla - iA - h\,\frac{\nabla u_o}{u_o})\phi \right\|^2 = \langle P_A^\Omega \phi/\phi \rangle - \lambda_o^\Omega \|\phi\|^2$$

From this identity, we recover first the classical lemma which follows also from the Kato's inequality (see [KA] or the survey [HU]) :

lemma 7.2.12 $\quad \lambda_A^\Omega \geqslant \lambda_o^\Omega$

Proof of lemma 7.2.2

This is a consequence of the minimax principle : $\lambda_A^\Omega = \inf\limits_{\phi \in C_o^\infty(\Omega)} \dfrac{\langle P_A^\Omega \phi/\phi \rangle}{\|\phi\|^2}$,

$$\lambda_o^\Omega = \inf_{\phi \in C_o^\infty(\Omega)} \frac{\langle P_o^\Omega \phi/\phi \rangle}{\|\phi\|^2}$$

(Note that $C_o^\infty(\Omega)$ is dense in the domain of $q_A$ and $q_o$)

#

We can now give the :

Proof of (i) $\Longrightarrow$ (iii)

Let us now suppose that : $\lambda_A^\Omega(h) = \lambda_o^\Omega(h)$ and let $u_A$ be a normalized eingenfunction of $P_A^\Omega(h)$ in $L^2(\Omega)$ corresponding to $\lambda_A^\Omega(h)$. Using the minimax principle and the compactness of $D(P_A^\Omega(h)$ in $L^2(\Omega)$ , we get (see Reed-Simon IV, Vol XIII. 15) a sequence $\phi_n \in C_o^\infty(\Omega)$ s.t

$$(7.2.1.2) \quad \begin{cases} \phi_n \longrightarrow u_A \text{ in } L^2(\Omega) \\ q_A(\phi_n,\phi_n) = (P_A^\Omega \phi_n/\phi_n) \longrightarrow \lambda_A^\Omega = \lambda_o^\Omega \quad \text{as } n \to \infty \end{cases}$$

Then we get from (7.2.1.1) and (7.2.1.2) that :

$$(7.2.1.3) \quad \left(h\nabla - iA - h\frac{\nabla u_o}{u_o}\right)\phi_n \longrightarrow 0 \text{ in } L^2(\Omega)$$

and

$$(7.2.1.4) \quad (h\nabla - iA - h\frac{\nabla u_o}{u_o})\, u_A = 0 \text{ in } \mathcal{D}'(\Omega) \text{ (and in } C^\infty(\Omega) \text{ because } u_A \in C^\infty(\Omega)).$$

Let : $\phi_A = u_A / u_o$ . We deduce from (7.2.1.4) the equation :

$$\text{(7.2.1.5)} \qquad h d \phi_A = i \ \phi_A . \ \omega_A$$

We prove now that :

$$\text{(7.2.1.6)} \qquad \phi_A \neq 0 \text{ in } \Omega.$$

Suppose indeed that $\phi_A(y_o) = 0$ for some $y_o \in \Omega$ . We deduce from (7.2.1.5) the inequality :

$$\text{(7.2.1.7)} \qquad |\nabla \phi_A| \leq C \ |\phi_A|$$

In a ball $B(y_o, r) \subset \Omega$ , we write :

$$\text{(7.2.1.8)} \qquad \phi_A(x) = \phi_A(y_o) + \int_o^1 \frac{d}{dt} \left(\phi_A(y_o + t(x-y_o)\right) dt$$

and from (7.2.1.7) and (7.2.1.8) we deduce :

$$\sup_{x \in B(y_o,r)} |\phi_A| < Cr \sup_{x \in B(y_o,r)} |\phi_A|$$

from which we get that

$$\phi_A(x) = 0 \qquad \text{in } B(y_o, \inf (r, \tfrac{1}{C}))$$

So $\phi_A^{-1}(0)$ is open in $\Omega$ and clearly closed in $\Omega$ . From the connexity of $\Omega$ ,we deduce : $\phi_A = 0$ which in contradiction with $\|u_A\| = 1$. (7.2.1.6) is proved.

From (7.2.1.5) , by taking the differential, we get :

$$0 = i \ \ d \phi_A \ \wedge \omega_A + i\phi_A \ \ \sigma_B = \frac{i}{h} \ \phi_A \, \omega_A \wedge \omega_A \qquad + i \ \phi_A \ \ \sigma_B$$

$$0 = \phi_A . \ \sigma_B$$

and finally :

$$\sigma_B = 0 \qquad\qquad \text{(this is } (iii)_b) \text{ (using 7.2.1.6)}$$

To get $(iii)_b$ , we observe, that, locally, for some choice of the logarithm, we can write (7.2.1.5) in the form :

$$\text{(7.2.1.9)} \qquad h \ d \ \text{Log} \ |\phi_A| + i \ h \ d \ (\text{Arg} \ \phi_A) = i \ \ \omega_A$$

which gives $\qquad |\phi_A| = \text{cst}$ locally

$$h d(\text{Arg} \ \phi_A) = \omega_A$$

The argument is defined modulo $2\pi$ so along a closed curve $\gamma$ in $\Omega$, we must have

$$\frac{1}{2\pi h}\int_{\gamma} \omega_A \in \mathbb{Z}.$$

7.2.2 the semi-classical approach $(h \to 0)$

A model on $S^1$

Let us first look to the following simple (but enlightening example) on $S^1$ :

(7.2.2.1) $\sum_j (h D_{x_j} - A)^2 + V$

with $A$ = cst and $V$ $C^\infty$.

If necessary we identify $V$ on $S^1$ with its natural extension on $\mathbb{R}$ as a $2\pi$-periodic function.

As a first exercise, let us take $V = 0$. Then by considering the Fourier series, we find immediatly that the spectrum is given by : $(h k - A)^2$ and we recover immediately the analog of theorem 7.2.1.1 saying :

$\lambda_A = \lambda_0$ if and only if $h k = A$ for some $k \in \mathbb{Z}$

(note the magnetic field is 0)

In the case where $V$ is not 0 but has a unique non degenerate minimum on $S^1$, after the change of function :

$u = e^{-iA\frac{x}{h}}.v$

We get the floquet problem studied in section 6 :

$$\text{(7.2.2.2)}\quad \begin{cases} ((h^2 D_x^2) + V)\, v = \lambda_A^\theta\, v \\ v(x+2\pi) = e^{i\theta} v(x) \qquad \text{with } \theta = 2\pi A/h \pmod{2\pi\mathbb{Z}} \end{cases}$$

Then using Theorem (6.2.4) with $\theta = 0$ and $\theta = 2\pi A/h$ we get :

Theorem 7.2.2.1 Under the hypotheses preceding (7.2.2.2) we have :

(7.2.2.3) $$\lambda_A(h) - \lambda_0(h) = \left[1 - \cos 2\pi \frac{A}{h}\right] h^{1/2}\, |a(h)|\, e^{-S_0/h} + O\left(e^{-S_0/h - \varepsilon_0/h}\right)$$

with $\varepsilon_0 > 0$, $O$ uniform with respect to $A$,

$$S_0 = \int_0^{2\pi} \sqrt{(V - \min V)(t)}\, dt$$

$a(h)$ is the realization of $\left(\sum_{j\in\mathbb{N}} a_j h^j\right)$ with $a_0 > 0$

A model in $R^2 \setminus B(0,\delta)$

Let us take $\Omega = R^2 \setminus B(0,\delta)$ for some $0<\delta<1$. We suppose that :

(7.2.2.4) $\vec{A}$ and $V \in C^\infty(\overline{\Omega})$

(7.2.2.5) $B = 0$ in $\overline{\Omega}$

(7.2.2.6) $V \underset{|x|\to\infty}{\longrightarrow} \infty$

(7.2.2.7) V has a unique non degenerate minimum in $\overline{\Omega}$ at the point $y = (1,0)$ and $V(y) = 0$

To simplify the proof, we make also the unessential hypothesis that :

(7.2.2.8) $V(x_1, -x_2) = V(x_1,x_2)$

We are interested in computing explicity the asymptotic as $h \longrightarrow 0$ of $\lambda^{\Omega}_{tA}(h) - \lambda^{\Omega}_{0}(h)$ which is strictly positive when $t\Phi \overset{\text{def}}{\equiv} \frac{t}{2\pi h}\int_{\gamma_0} A \notin Z$ (where $\gamma_0 = \partial B(0,\delta)$) by theorem 7.2.1.1 and lemma 7.2.1.2

Let us introduce

(7.2.2.9) $S_0 = d_V(y,\partial\Omega)$

and

(7.2.2.10) $S_1 = \inf_{\gamma\in\Gamma}$ lengh $\gamma$

where $\Gamma$ is the set of the closed paths in $\Omega$ with $y\in\gamma$ which are not homotop to 0 and the lenght is computed with respect to the Agmon metric in $\Omega$.

The fundamental hypothesis (which will say that the effect of the boundary is weak because V creates a barrier between y and $\partial\Omega$) is :

(7.2.2.11) $S_1 < 2\, S_0$

To simplify, we shall also assume :

(7.2.2.12) the minimum $S_1$ is obtained along a unique path $\gamma_1$ and this path is non-degenerate in the classical sense (see 4.4.42)

This geometrical condition is translated analytically by :

(7.2.2.13) $\left(\frac{\partial^2 d_V}{\partial x_1^2}\right)(z)>0$ where $d_V = d_V[x,y]$

at the point z corresponding to the intersection of $\gamma_1$ and $x_2 = 0$, $x_1 < 0$

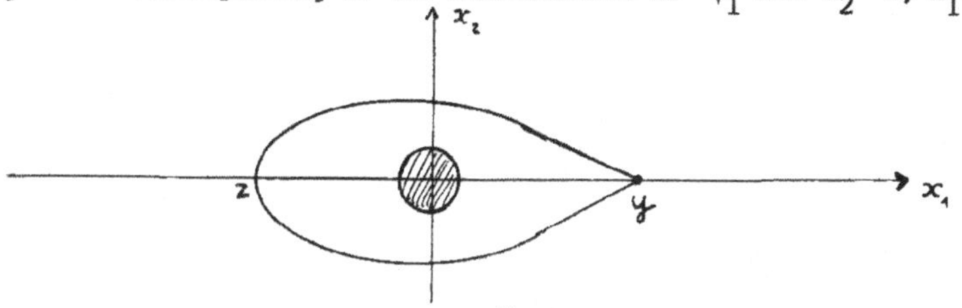


Then we have the following theorem (cf $[HE]_3$)

Theorem 7.2.2.2

Under hypothesis (7.2.2.4) ——— (7.2.2.13) , we have :

(7.2.2.14) $\lambda^{\Omega}_{tA}(h) - \lambda^{\Omega}_{o}(h) = h^{1/2}\left(1 - \cos\frac{t\Phi}{h}\right) a(h)\, e^{-S_1/h} + O(e^{-(S_1+\varepsilon_o)/h})$

where $\varepsilon_o > 0$ , O is uniform with respect to t

a(h) is the realization of $\sum_{j\in N} a_j h^j$ with $a_o > 0$

Proof of the theorem

In the universal covering $\Omega^R$ of $\Omega$ , we introduce the Agmon distance $d_V^R$ associated to the potential $V^R$ defined by $V^R(x) = V(\Pi(x))$ where $\Pi$ is the projection of $\Omega^R$ onto $\Omega$ .

For $\varepsilon_1 > 0$ , let us denote by $M_{\varepsilon_1}$ the open set defined in $\Omega_R$ by :

$$M_{\varepsilon_1} = \{\dot{x} \in \Omega_R,\ \dot{x} = \rho e^{i\theta},\quad \rho \in ]\delta, +\infty[,\ \theta \in ]-\Pi-\varepsilon_1, \Pi+\varepsilon_1[\}$$

We take here the natural global polar coordinates in $\Omega^R$ and the point y is identify with the point in $\Omega^R$ : $\rho = 1, \theta = 0$.

This explicit choice of $M_{\varepsilon_1}$ is possible due to (7.2.2.8), but the important property is that $M_{\varepsilon_1}$ contains a ball $B\left(y, \frac{S_1}{2} + \eta\right)$ for some $\eta > 0$ in $\Omega^R$ (the ball is for the Agmon distance $d_V^R$).

Our reference problem will be the Dirichlet realization of the operator $P_o(h)$ in $M_{\varepsilon_1}$ : $P_o^{M_{\varepsilon_1}}(h)$

Let $u_o^R$ be the first eigenfunction of $P_o^{M_{\varepsilon_1}}(h)$ associated to $\lambda_o^R(h)$.

Let us recall the properties of $u_o^R$ proved in § 2-4 which will be used later on ;

$$(7.2.2.15) \quad u_o^R(x) = O_\varepsilon\left(e^{-(1-\varepsilon)\frac{d^R(\dot{x},y)}{h}+\frac{\varepsilon}{h}}\right), \ \forall\ \varepsilon>0$$

Let us also remark that $\gamma_1$ can be seen as the projection of the minimal geodesics

$\gamma_1^{(1)}$ (resp. $\gamma_1^{(-1)}$) between y and the point $y^{(1)}$ corresponding to ($\rho=1$, $\theta=2\pi$) (resp. $y^{(-1)}$ corresponding to $\rho=1$, $\theta=-2\pi$)

It is then clear that $u_o^R$ has a good B.K.W approximation in $M_{\varepsilon_{1/2}}$ in a neighborhood of $\gamma_1^{(1)}$ or $\gamma_1^{(-1)}$, in particular in the neighborhood of the points $z^{(1)}$ and $z^{(-1)}$ s.t $\Pi(z^{(j)}) = z$ ($j = \pm 1$) of the form : $a(x,h)\, e^{-\frac{d^R(\dot{x},y)}{h}}$ (See section 4.4)

Let $A^R(\dot{x})$ the natural magnetic potential one-form on $\Omega^R$:

$$(7.2.2.16) \quad \vec{A}^R(x) = \vec{A}(\Pi(\dot{x}))$$

and $\phi^R$ the solution in $\Omega^R$ of :

$$(7.2.2.17) \quad \phi^R(y) = 0\ , \qquad \nabla\phi^R = \vec{A}^R$$

Let $\chi^R$ a $C^\infty$ function on $\Omega^R$ with compact support in $M_{\varepsilon_1}$ and equal to 1 on the ball $B(y, \frac{S_1}{2}+\eta)$ for some $\eta>0$.

The function $u_{tA}$ defined below is the good candidate to give an approximate eigenfunction for $P_{tA}^\Omega(h)$ . We define :

$$(7.2.2.18) \quad u_{tA}(x) = \sum_{\substack{\Pi(\dot{x}) = x \\ \dot{x}\in\Omega^R}} \chi^R(x)\, e^{\frac{it\phi^R(\dot{x})}{h}}\, u_o^R(\dot{x})$$

clearly $u_{tA}$ is a $C^\infty$ function on $\Omega$ , in the domain of $P_{tA}^\Omega$ and it is clear that the sum appearing in the r.h.s of 7.2.2.18 is finite.

Let us compute now :

$$(7.2.2.19) \quad (P_{tA}^\Omega\, u_{tA})\,(x) = \left(\sum_{\substack{\pi(\dot{x}) = x \\ \dot{x}\in\Omega^R}} e^{it\phi_h^R(\dot{x})/h}\, ([P,\chi^R]\, u_o^R)\,(\dot{x})\right) + \lambda_o^R\, u_{tA}(x)$$

According to (7.2.2.15) and to the choice of $\chi^R$ it is clear that there exists an eigenvalue $\lambda_{tA}^{\Omega}$ s.t

$$\lambda_{tA}^{\Omega}(h) - \lambda_{o}^{R}(h) = O(e^{-(\frac{S_1}{2}+\eta)/h}) \tag{7.2.2.20}$$

The end of the proof is quite analogous to the discussion in section 6 and we omit some details.

To compute more precisely $\lambda_{tA}^{\Omega}(h) - \lambda_{o}^{R}(h)$, we have to compute :

$((P_{tA}^{\Omega} - \lambda_o^R)u_{tA}/u_{tA})$ which is equal to $\lambda_{tA}^{\Omega} - \lambda_o^R(h)$ modulo $(O(e^{-\frac{S_1+\eta}{h}}))$ (see lemma 6.2.3 or § 4.3)

Noting that $\phi^R(z^{(1)}) - \phi^R(z^{(-1)}) = \int_\gamma \omega_A^R = \Phi$, we get :

$$\lambda_{tA}^{\Omega}(h) - \lambda_o^R(h) = e^{it\Phi/h}\rho_1(h) + e^{-it\Phi/h}\rho_{(-1)}(h) + O(e^{-\frac{S_1+\eta}{h}}) \tag{7.2.2.21}$$

where $\rho_1(h) = \int ([P,\chi^R]u_o^R)(\Pi_1^{-1}(x)).\ u_o^R(\Pi_{(-1)}^{-1}(x))\,dx$

$\rho_{(-1)}(h) = \int ([P,\chi^R]u_o^R)(\Pi_{(-1)}^{-1}(x))\ .\ u_o^R(\Pi_1^{-1}(x))\,dx$

where

$(\Pi_1^{-1})$ satisfies to: $\Pi \circ \Pi_1^{-1} = id$, $\quad \Pi_1^{-1}(z) = z^{(1)}$ and

$(\Pi_{(-1)}^{-1})$ to $\Pi\circ\Pi_{(-1)}^{-1} = id$ , $\quad \Pi_{(-1)}^{-1}(z) = z^{(-1)}$

$\rho_1$ and $\rho_{-1}$ are in fact equal (cf § 6) and appear as interaction coefficients between the wells $y$ and $y^{(1)}$ or $y$ and $y^{(-1)}$.

According to the B.K.W approximation and the Stokes formula which reduces the computation of $\rho_1(h)$ to an integral on $x_2 = 0$, $x_1 < 0$ (see § 4.4) in the neighborhood of $z$, we get using (7.2.2.13) the announced expansion :

$\lambda_{tA}^{\Omega}(h) - \lambda_o^R(h) = -h^{1/2}(\cos t\frac{\Phi}{h})\,a(h) + O(e^{-(S_1+\varepsilon_o)/h})$

and finally (7.2.2.14) by difference (note that a(h) is independant of t)

Remark 7.2.2.3

The effect which appears here is related to the classical Aharonov-Bohm effect (See the discussion in [HE]$_3$)

The magnetic field is 0 in $\Omega$ but some effect remains due to the fact that $\frac{1}{2\pi h}\int_\gamma A = t\Phi/_{2\pi h} \notin Z$.

This is a purely quantum effect ; Note however that the classical Aharonov-Bohm effect [AH-BO] is more a problem of scattering.

The physicists could say that the Dirichlet condition (which corresponds to an infinite barrier) is not realistic. Let us finish this subsection with a small perturbation of the example in $\Omega$ but now in $R^n$.

The magnetic flea

This is the "magnetic" analog of an effect first studied by Jona-Lasinio Martinelli-Scoppola [J.M.S] and then in [HE-SJ]$_2$ called by B.Simon [SI]$_5$ : the flea of the elephant. The theorem is the following

Theorem 7.2.2.4 Let

(7.2.2.22) $\vec{A}, V \in C^\infty(R^2)$ , $V \underset{|x|\to\infty}{\longrightarrow} \infty$

(7.2.2.22) Supp B C B (0,$\delta$)

(7.2.2.23) V has a unique non degenerate minimum in $R^n$ at the point y = (1,0) and V(y) = 0

Suppose moreover that (7.2.2.8) - (7.2.2.13) are satisfied, then the first eigenvalue of $P_{tA}(h)$ in $R^2$ satisfies to :

(7.2.2.24) $\lambda_{tA}(h) - \lambda_o(h) = (1-\cos\frac{t\Phi}{h})\, a(h)\, e^{-S_1/h} + O(e^{-(S_1+\varepsilon_o)/h})$

with the same conventions as in Theorem 7.2.2.2

Proof We have just to compute the difference $\lambda^\Omega_{tA}(h) - \lambda_{tA}(h)$ of the first eigenvalues of 2 Dirichlet problems.

In the case t=0, we have seen in § 4 (Remark 4.3.3) that this error is like

$$O_\varepsilon\left(e^{-\frac{d(y,\partial\Omega)}{h}+\varepsilon/h}\right) \quad (\forall \varepsilon>0)$$

If you analyze the proof, we have also the same result for $t \neq 0$, if we prove that the eigenfunctions of $P_{tA}$ or $P^{\Omega}_{tA}$ decrease near the bottom like in the case $t=0$. This property is indeed true (cf [HE-SJ]$_{10}$) and can be deduced as in §3-3, once we have the analog of the theorem 3.1.1 with magnetic fields (cf also Proposition 6.2.2) :

Proposition 7.2.25

Let $\Omega$ an open domain with $C^2$ boundary. Let $V, A \in C^\infty(\bar{\Omega})$.
Then, if $\phi$ is a Lipschitzien function on $\Omega$, we have, for any u in $C^2_o(\bar{\Omega},\mathbb{C})$ with $u_{/\partial\Omega} = 0$ ($C^2_o(\bar{\Omega};\mathbb{C})$ is the space of restrictions to $\Omega$ of functions of $C^2_o(R^n)$)

$$\int_\Omega \left|\nabla_A^{(h)}(e^{\phi/h}u)\right|^2 dx + \int_\Omega (V-|\nabla\phi|^2)\, e^{2\phi/h} |u|^2\, dx$$

$$= \mathrm{Re} \int_\Omega e^{2\phi/h}\, P_A u . \bar{u}\, d\,x$$

with $\nabla^{(h)}_A = h\nabla - iA$

## § 7.3 On the multiplicity of the first eigenvalue

It is well known that the multiplicity of the first eigenvalue of the Schrödinger operator (case A=0) is simple and that the first eigenfunction can be chosen strictly positive. Examples where the multiplicity is 2 have been given in [A.H.S] , [LA-O'CA], $[HE\text{-}SJ]_{10}$ , $[HE]_3$ . We present here one example from $[HE]_3$ and a variation of an example of [LA-O'CA].

Example 1 We return to the situation described in § 7.2 where $\Omega = R^2 \setminus B(0,\delta)$ $(\delta > 0)$

We make the following assumptions :

(7.3.1) $V>0$ ; $V \in C^\infty(\Omega)$ , $V(x) = V(-x)$ in $\bar{\Omega}$ , $V \underset{|x|\to\infty}{\longrightarrow} \infty$

(7.3.2) $\vec{A}(-x) = -\vec{A}(x)$ , $B(x) = 0$ , $|A| \leq CV^{1/2}$ +

(We can take for example $\omega_A = \dfrac{-x_2\, dx_1 + x_1\, dx_2}{x_1^2 + x_2^2}$)

Let us also introduce the operator $\Sigma$ by :

(7.3.3) $(\Sigma u)(x) = u(-x)$

According to (7.3.1) and (7.3.2) , $\Sigma$ commutes with $P^\Omega_{tA}$ for each t and we can decompose the spectrum into an odd spectrum and an even spectrum. To prove that for some value of t the first eigenvalue is of multiplicity 2 it is sufficient to find values $t_o, t_1$ $(t_o \neq t_1)$ for which we have

$$\lambda^{\Omega\ \text{even}}_{t_o} \leq \lambda^{\Omega\ \text{odd}}_{t_o} \quad (\text{and } \lambda^{\Omega\ \text{even}}_{t_1} \geq \lambda^{\Omega\ \text{odd}}_{t_1})$$

and from the minimax principle we get the continuity with respect to t of $\lambda^{\Omega\ \text{even}}_t$ or $\lambda^{\Omega\text{odd}}_t$ which are the first even (or odd) eigenvalues of $P^\Omega_{tA}$.

Then there exists $t_2$ with :

(7.3.4) $\lambda^{\Omega\ \text{even}}_{t_2} = \lambda^{\Omega\ \text{odd}}_{t_2}$

Let us take $t_o = 0$. It is classical that:

(7.3.5) $\lambda^{\Omega\text{even}}_o < \lambda^{\Omega\ \text{odd}}_o$

Let $\phi^R$ a solution of $\nabla\phi^R = A^R$ in $\Omega^R$

For $t = \frac{2\Pi}{\Phi}$ , $e^{it\phi^R(x)}$ is well defined in $\Omega$ and in particular : $P_{2\pi A/\Phi}$ , is unitary equivalent to $P_o$ by conjugation through $e^{i2\Pi\phi^R/\Phi}$

Let us observe now that $x \longrightarrow e^{-i\, 2\Pi\, \phi^R(x)/\Phi}$

is odd because of the relation :

$$\frac{2\Pi}{\Phi}\,(\phi^R(x) - \phi^R(-x)) = \frac{2\Pi}{\phi}\,\left(\frac{1}{2}\int_\gamma \omega_A\right) = \Pi$$

Then we get that the first eigenfunction of $P_{2\Pi A/\Phi}$ is odd :

(7.3.6) $\lambda_o^{\Omega\text{ even}} = \lambda_{2\Pi/\Phi}^{\Omega\text{ odd}} < \lambda_{2\Pi/\Phi}^{\Omega\text{ even}} = \lambda_o^{\Omega\text{ odd}}$

<u>Remark 7.3.1</u>

If we want an example in $R^2$ we observe that this argument is stable by perturbation. In the spirit of §7.2.2 , we can semi classically minorize

$\lambda_o^{\Omega\text{ odd}}(h) - \lambda_o^{\Omega\text{ even}}(h)$ and estimate $\lambda_{tA}^{\Omega\text{ odd}}(h) - \lambda_{tA}^{R^2\text{odd}}$ , $\lambda_{tA}^{\Omega\text{ even}}(h) - \lambda_{tA}^{R^2\text{even}}(h)$

and we get an example on $R^2$ (cf [HE-SJ].[HE]$_3$).

<u>Remark 7.3.2</u>

In fact, it is possible to prove that it is true for $t_2 = \Pi/\phi$.
Let us consider the operator J defined by :

(7.3.7) $(Ju)(x) = e^{-2i\Pi\phi^R/\Phi}\,\bar{u}(x)$

J commutes with $P^\Omega_{\Pi A/\Phi}$ , is antilinear, and verifies $K^2 = -I$

Consider K :

(7.3.8) $K = \int J$

Then

(7.3.9) K commutes with $P^\Omega_{\Pi A/\Phi}$ , is antilinear, and verifies $K^2 = -I$

As in [LA-LI] , the Kramer's theorem gives that the multiplicity is at least 2 for <u>all</u> eigenvalues.

Indeed if u is a non zero eigenvector , Ku and u are linearly independant. This is easily proved as follows. If $Ku = \lambda u$ then :

$K^2u = -u = K(\lambda u) = \bar{\lambda}Ku = |\lambda|^2u$ (contradiction)

Example 2

Let us describe now a variant of the examples of [A.H.S] , [IA. O'CA] . We consider the following family of Schrödinger operators :

$$P_t(h) = (h\,D_{x_1} - t\,x_2)^2 + (hD_{x_2} + tx_1)^2 - (x_1^2 + x_2^2) + (x_1^2 + x_2^2)^4$$

For t = 0, it is the classical Schrödinger operator. The first eigenfunction is of multiplicity 1 and is invariant by rotation.

In fact, we can decompose $L^2(R^2)$ as a direct hilbertian sum :

7.3.10 $$L^2(R^2) = \sum_{m\in Z}^{\oplus} E_m$$

where $E_m = \{u \in L^2(R^2)\ ,\ R_\theta u = e^{im\theta}\,u,\ \ \theta \in R\}$

(where $R_\theta$ is the rotation of angle $\theta$ )

Because $R_\theta$ commutes with $P_t(h)$ for each $\theta$ , we can write :

7.3.11 $$P_t(h) = \sum_{m\in Z}^{\oplus} (P_{t\wedge E_m})$$

Let us denote by $\lambda_o^m(t)$ the first eigenvalue of $P_t^m = P_{t\wedge E_m}$

As in example 1, to get the existence of a first double eigenvalue, it is sufficient to find $t_1>0$ and some $m \in Z\setminus 0$ s.t :

7.3.12 $$\lambda_o^m(t_1) < \lambda_o^o(t_1)$$

Then we shall get the existence of some $t_2>0$ and some $m_o \in Z\setminus 0$

s.t ⓐ $\lambda_o^o(t) < \lambda_o^m(t)$ $\quad t \in [0,t_2[$, $m \in Z\setminus 0$

(7.3.13) ⓑ $\lambda_o^o(t_2) = \lambda_o^{m_o}(t_2)$

To prove that (7.3.12) $\Longrightarrow$ (7.3.13), let us observe that $t \to \lambda_o^m(t)$ is continuous with respect to t.

If (7.3.13) was not true (because it is clear that (7.3.13)ⓐ is satisfied for $t\in[0,t_o]$ (with $t_o>0$), $\forall m \in Z\setminus 0$) it will imply that for each $m \in Z\setminus 0$, for each $t> 0$

$\lambda_o^o(t) < \lambda_o^m(t)$

is contradiction with (7.3.12) . (Note that $\inf_{m \in \mathbb{Z} \setminus 0} \lambda_o^m(t)$ is continuous, because for some $t_0$ the minimum is obtained only for a finite set of m because the $\lambda_o^m(t)$ belong to the spectrum of an operator $P_t$ with compact resolvent).

Let us prove now (7.3.12) for m=-1.

The study of $P_t$ on $E_m$ is equivalent to the study of

$$(hD_{x_1})^2 + (hD_{x_2})^2 + (t^2-1)\ (x_1^2 + x_2^2) \ + th\,m + (x_1^2 + x_2^2)^4$$

Then, we get by semi-classical arguments (using the harmonic approximation) (see § 2.1) and for t>1 :

$$\lambda_o^o(h;t) = (2\ \sqrt{t^2-1}\ )\ h + O(h^{3/2})$$

$$\lambda_o^{-1}(h;t) = (-t+ 4\sqrt{t^2-1}\ )h + O\ (h^{3/2})$$

Finally $\lambda_o^o(h,t) - \lambda_o^{-1}(h,t) = (-2\sqrt{t^2-1} + t)\ h + O(h^{3/2})$

For t>1 and not far from 1 and h small enough we get (7.3.12).

#

REFERENCES.


[AB.MA] R.Abraham-J.Marsden :
Foundations of mechanics , second edition

[AB.RO] R.Abraham-J.Robbin :
Transversal mapping and flows (Appendix C by Al.Kelley)
W.A.Benjamin ,Inc.

[AG] S.Agmon :
Lectures on exponential decay of solutions of second order elliptic equations .Bounds on eigenfunctions of N-Body Schrodinger operators
Mathematical notes of Princeton university

[AH.BO]Y. Aharonov-D.Bohm :
Significance of Electromagnetic Potentials in the quantum theory
Phys.Rev.Vol.115 , n°3 ,Aug 1959

[AR] V.Arnold :
Mathematical methods of mechanics , Springer verlag

[A.H.S]J.Avron-I.Herbst-B.Simon :
[1]Schrodinger operators with magnetic fields
I General interactions
Duke Math. Journal 45 (1978) , 847-884
[2]Schrodinger operators with magnetic fields
II Separation of the center of mass in Homogeneous magnetic fields
Ann.Phys.114(1978) , 431-451
[3]Schrodinger operators with magnetic fields
III Atoms in Homogeneous Magnetic fields
Commun.Math.Phys.79,529-572 (1981)

[BI] J.M.Bismut :
The Witten complex and the degenerate Morse inequalities
Journal of differential equations
Vol23,n°3,May1986 p207-240

[BO] R.Bott :
Lectures on Morse theory , old and new
Bull.of the A.M.S Vol.7 n°2,sept.1982,p.331-358

[BR] F.Bruhat :
"Travaux de Sternberg "
Séminaire Bourbaki 60-61 ,exposé n°217

[CA] U.Carlsson : In preparation

[CO] S.Coleman :
The uses of instantons
Proc.Internat.School of Physics , Erice , 1977

[CDV.] Y.Colin de Verdière :
[1] Spectre conjoint d'opérateurs qui commutent
Duke Math.J.46, 1979,p.169,182
[2]L'asymptotique de Weyl pour des bouteilles magnétiques
Comm.in Math.Physics 105 (1986) p 327-335

[C.D.S] J.M.Combes-P.Duclos-R.Seiler :
[1]Krein's formula and one dimensional multiple well
J.of functional Analysis 52 (1983) p.257-301
[2]Convergent expansions for tunneling
Comm.in Math.Phys.92 (1983) p.229-245

[C.S.S] J.M.Combes-R.Schrader-R.Seiler :
classical bounds and limits for distributions of Hamiltonian operators in electromagnetic fields
Ann.of Physics 111 (1978) p.1-18

[C.F.K.S] H.L.Cycon,R.G.Froese,W.Kirsch,B.Simon :
Schrödinger operators
(with applications to quantum mechanics and global geometry)
Texts and Monographs in Physics,Springer Verlag

[DE] J.P.Demailly :
Champs magnétique et inégalités de Morse pour la d"-cohomologie
Annales de l'institut Fourier 35 ,189-229 (1985)

[DU] A.Dufresnoy :
Un exemple de champ magnétique dans $R^n$
Duke Math.Journal 53(3),1983, p729-734

[EA] M.S.P.Eastham :
The spectral theory of Periodic Differential Equations
Hafner,New york,1974

[ELS.WA] A.El Soufi-X.P.Wang :
Quelques remarques sur la méthode de Witten : cas du théorème de Poincaré-Hopf et d'une formule d'Atiyah-Bott
Publications internes de l'institut Fourier (1986)
(to appear in annals of global Analysis and Geometry )

[FE.MA] Fedoryuk-V.P.Maslov :
Semi-classical approximation in Quantum Mechanics
Reidel 1981

[HA] E.Harrell :
[1] On the rate of asymptotic eigenvalue degeneracy
Comm.in Math.Phys. 60(1978) , 73-95
[2]The band structure of a one dimensional periodic system in the scaling limit
Ann.of Physics 119 (1979) p.351-369
[3] Double Wells
Comm.math. Physics75 (1980) , 239-261

[HE] B.Helffer :
[1] Théorie spectrale pour des opérateurs globalement elliptiques
Astérisque n°112
[2] Partial differential equations on nilpotent groups
conférences à l'université de Maryland
Lecture notes in Mathematics n°1077, p.210-254
[3] Sur l'équation de Schrodinger avec champ magnétique
Exposé au séminaire EDP de l'école Polytechnique
[4]Etude du Laplacien de Witten associé à une fonction de Morse dégénérée
Publications de l'université de Nantes ,Séminaire EDP 85-86

[HE.MO] B.Helffer, A.Mohamed :
Sur le spectre essentiel des opérateurs de Schrodinger avec champ magnétique
To appear Ann.Institut Fourier (1988)

[HE.NO] B.Helffer,J.Nourrigat :
Hypoellipticité maximale pour des opérateurs polynômes de champs de vecteur
Progress in Mathematics Birkhauser vol.58

[HE.RO] B.Helffer,D.Robert :
[1]Comportement semi-classique du spectre des hamiltoniens quantiques elliptiques
Annales de l'institut Fourier 31(3)(1981) p.169-223
[2]Comportement semi-classique du spectre des hamiltoniens quantiques hypoelliptiques
Annales de l'ENS de Pise Série IV , Vol IX , n°3(1982)
[3]Calcul fonctionnel par la transformée de Mellin et applications
Journal of functional Analysis , Vol.53, n°3, oct.1983
[4]Puits de potentiel généralisés et asymptotique semi-classique
Annales de l'IHP (Physiquethéorique), Vol.41,n°3, 1984,p..291-331

[HE.SJ] B.Helffer,J.Sjostrand :
[1] Multiple wells in the semi-classical limit I
comm.in PDE , 9(4), p.337-408 (1984)
(annoncé aux actes du colloque de saint Jean de Monts en Juin 1983)
[2] Puits multiples en limite semi-classique II
-Interaction moléculaire-Symétries-Perturbations
Annales de l'IHP ( Physique théorique ) ,Vol.42,n°2,1985,p.127-212
[3] Multiple wells in the semi-classical limit III
Math.Nachrichte 124(1985) p.263-313
[4]Puits multiples en limite semi-classique IV
-Etude du complexe de Witten -
Comm.in PDE Vol.10 , n°3 , 1985 , p.245-340
[5] Puits multiples en limite semiclassique V
le cas des minipuits -
Volume in honor of S.Mizohata 1986
(annoncé au séminaire de Nantes 1984-1985 et au séminaire de l'école Polytechnique exposé n°10 1985-86 )
[6]Effet tunnel pour l'opérateur de Schrodinger semi-classique I
Série de conférences aux journées EDP de Saint-Jean de Monts (Juin 1985)
[7]Effet tunnel pour l'opérateur de Schrodinger semi-classique II
Advances in microlocal Analysis P.291-323
NATO Series - Mathematics Vol.168 D.Reidel
[8]Résonances en limite semi-classique
Mémoires de la SMF (1986) Tome 114 Fasc3
[9] Puits multiples en limite semi-classique VI
- le cas des puits variétés -
Annales de l'IHP ( physique théorique), vol 46,n°4 1987,p.353-372
[10]Effet tunnel pour l'équation de Schrodinger avec champ magnétique
Preprint de l'école polytechnique (Dec.1986)
to appear in annales of the ENS of PISE
[11]A proof of the Bott inequalities
( soumis pour un volume en l'honneur de M.Sato )

[HEN] G.Henniart :
Les inégalités de Morse d'après Witten
Séminaire Bourbaki 36ème année (1983-1984) n°617

[HO] L.Hörmander :
On the Asymptotic distribution of eigenvalues of p.d.o in $R^n$
Arkiv för Math.17 n°3(1981) p.169-223

[HU] W.Hunziker :
Schrôdinger operators with Electric or Magnetic fields
Proc.Int.Conf.in Math.Phys., Lausanne
Lecture Notes in Physics 116 (1980)

[IV] Y.V.Ivrii :
[1] On quasi-classical spectral asymptotics for the Schrödinger equation
Dokl.Akad.Nauk. SSSR (1982)
[2]Les estimations pour le nombre de valeurs propres négatives de l'opérateur de Schrödinger
CRAS (1986) t.302 n°13,14,15 pp467-470,491-494,535-538
[3] Estimates for a number of negative eigenvalues of the Schrödinger operator with intensive magnetic field
Proc.of St Jean de Monts juin 1987

[IW] A.Iwatsuka :
[1] The essential spectrum of two-dimensional schrödinger operators with perturbed magnetic fields
J.Math.Kyoto Univ. 23(3) (1983) p475-480
[2] Magnetic Schrödinger Operators with compact resolvent
J.Math.Kyoto Univ. 26(3) (1986) p357-374

[J.M.S] G.Jona-Lasinio,F.Martinelli,E.Scoppola :
New approach to the semiclassical limit of quantum mechanics ,
I multiple tunneling in one dimension
Comm.Math.Phys. 80 (1981) , 223

[KA] T.Kato :
[1]Perturbation theory for linear Operators
Springer verlag 1966
[2] Israel Journal of Math. 13 (1972) p125-174

[KO] J.J.Kohn :
Lecture on degenerate elliptic problems
CIME1977 ,p 91-149

[LA.LI]L.D.Landau-E.M.Lifshitz :
Mécanique quantique,Théorie non-relativiste
Editions Mir (1974)

[LA.O'CA] R.Lavine et M.O'Carroll :
Ground state properties and lower bounds on energy levels of a particle in a uniform magnetic field and external potential
J.Math.Phys.18 (1977),1908-1912

[LE] J.Leray :
[1]Analyse Lagrangienne en Mécanique semi-classique
Cours au collège de France (1976-1977)

[MAR] A.Martinez :
[1]Estimations de l'effet tunnel pour le double puits
Journal de Math. pures et appliquées , T.86 Fasc.2 (1987)p.195
[2] Estimations de l'effet tunnel pour le double puits II
To appear in bulletin de la SMF 1987

[MAS] V.P.Maslov :
Théorie des perturbations et Méthodes asymptotiques
Dunod

[MELI] A.Melin :
Parametrixe constructions for some classes of right invariant differential operators on nilpotent groups
Annals of global Analysis and geometry 1 , 79-130 (83)

[MELR]R.Melrose :
Elliptic operators on Manifolds MIT 1984

[MI] J.Milnor :
Morse theory
Princeton university press (1963)

[MO] A.Mohamed :
Quelques remarques sur le spectre de l'opérateur de Schrödinger avec un champ magnétique
preprint 1987

[O.P] S.Olariu-I-Iowitzu Popescu :
The quantum effects of electromagnetic fluxes
Review of modern Physics vol.57 n°2 (1985)

[OU] A.Outassourt :
Analyse semiclassique pour des opérateurs de Schrodinger avec potentiel pèriodique
Journal of functional Analysis ,Vol.72,n°1,May 1987

[RE.SI] M.Reed-B.Simon :
Methods of modern Mathematical Physics
Academic press

[RO] D.Robert :
[1] Comportement asymptotique des valeurs propres d'opérateurs du type Schrodinger à potentiel dégénéré
J.Math.Pures Appl. 61 (1982)
[2] Calcul fonctionnel pour les opérateurs admissibles et applications
Journal of functional Analysis Vol.45 n°1 (1982)
[3]Autour de l'approximation semi-classique
Progress in Mathematics , Birkhauser (1986)

[SHU] M.Shubin :
Pseudodifferential operators and spectral theory
Nauka Moscow (1978) (in russian)

[SI] B.Simon :
[1] Some quantum operators with discrete Spectrum but classically continuous spectrum
Annals of Physics 146,209-220(1983)
[2]Non classical Eigenvalue Asymptotics
Journal of functional Analysis , vol.53 ,n°1 ,August 1983
[3] Instantons,double wells and large deviations
Bull.AMS 8 (1983),323-326,
[4] Semi-classical Analysis of low lying eigenvalues I
Non degenerate minima: Asymptotic expansions,
Ann.Inst.H.Poincaré 38,(1983) , p295-307
[5]Semi-classical Analysis of low lying eigenvalues II
Tunneling
Annals of Mathematics,120 (1984),89-118
[6] Semi-classical Analysis of low lying eigenvalues III
Width of the ground state band in strongly coupled solids
Ann. of Physics 158 (1984) p 415-420
[7] Semi-classical Analysis of low lying eigenvalues IV
The flea of the elephant
Journal of functional analysis vol.63,n°1 ,August 1985

[SJ] J.Sjôstrand:
Singularités analytiques microlocales
Astérisque n°95 (1982)

[TA] H.Tamura:
Asymptotic distribution of Eigenvalues för Schrödinger operators with magnetic fields
Nagoya Math.Journal vol.105 (1987) p.49-69

[WIL]C.Wilcox :
Theory of Bloch Waves ,
J.Analyse Math. 33(1978),146-167

[WIT] E.Witten :
Supersymmetry and Morse theory
J.Diff.Geom.17 (1982),661

[W.Y] T.T.Wu,C.N.Yang :
Concept of non-integrable factors and global formulation of gauge fields
Phys.Rev.D.Vol.12,n°12 (1985)

INDEX.

Vol. 1173: H. Delfs, M. Knebusch, Locally Semialgebraic Spaces. XVI, 329 pages. 1985.

Vol. 1174: Categories in Continuum Physics, Buffalo 1982. Seminar. Edited by F.W. Lawvere and S.H. Schanuel. V, 126 pages. 1986.

Vol. 1175: K. Mathiak, Valuations of Skew Fields and Projective Hjelmslev Spaces. VII, 116 pages. 1986.

Vol. 1176: R.R. Bruner, J.P. May, J.E. McClure, M. Steinberger, $H_\infty$ Ring Spectra and their Applications. VII, 388 pages. 1986.

Vol. 1177: Representation Theory I. Finite Dimensional Algebras. Proceedings, 1984. Edited by V. Dlab, P. Gabriel and G. Michler. XV, 340 pages. 1986.

Vol. 1178: Representation Theory II. Groups and Orders. Proceedings, 1984. Edited by V. Dlab, P. Gabriel and G. Michler. XV, 370 pages. 1986.

Vol. 1179: Shi J.-Y. The Kazhdan-Lusztig Cells in Certain Affine Weyl Groups. X, 307 pages. 1986.

Vol. 1180: R. Carmona, H. Kesten, J.B. Walsh, École d'Été de Probabilités de Saint-Flour XIV – 1984. Édité par P.L. Hennequin. X, 438 pages. 1986.

Vol. 1181: Buildings and the Geometry of Diagrams, Como 1984. Seminar. Edited by L. Rosati. VII, 277 pages. 1986.

Vol. 1182: S. Shelah, Around Classification Theory of Models. VII, 279 pages. 1986.

Vol. 1183: Algebra, Algebraic Topology and their Interactions. Proceedings, 1983. Edited by J.-E. Roos. XI, 396 pages. 1986.

Vol. 1184: W. Arendt, A. Grabosch, G. Greiner, U. Groh, H.P. Lotz, U. Moustakas, R. Nagel, F. Neubrander, U. Schlotterbeck, One-parameter Semigroups of Positive Operators. Edited by R. Nagel. X, 460 pages. 1986.

Vol. 1185: Group Theory, Beijing 1984. Proceedings. Edited by Tuan H.F. V, 403 pages. 1986.

Vol. 1186: Lyapunov Exponents. Proceedings, 1984. Edited by L. Arnold and V. Wihstutz. VI, 374 pages. 1986.

Vol. 1187: Y. Diers, Categories of Boolean Sheaves of Simple Algebras. VI, 168 pages. 1986.

Vol. 1188: Fonctions de Plusieurs Variables Complexes V. Séminaire, 1979–85. Edité par François Norguet. VI, 306 pages. 1986.

Vol. 1189: J. Lukeš, J. Malý, L. Zajíček, Fine Topology Methods in Real Analysis and Potential Theory. X, 472 pages. 1986.

Vol. 1190: Optimization and Related Fields. Proceedings, 1984. Edited by R. Conti, E. De Giorgi and F. Giannessi. VIII, 419 pages. 1986.

Vol. 1191: A.R. Its, V.Yu. Novokshenov, The Isomonodromic Deformation Method in the Theory of Painlevé Equations. IV, 313 pages. 1986.

Vol. 1192: Equadiff 6. Proceedings, 1985. Edited by J. Vosmansky and M. Zlámal. XXIII, 404 pages. 1986.

Vol. 1193: Geometrical and Statistical Aspects of Probability in Banach Spaces. Proceedings, 1985. Edited by X. Fernique, B. Heinkel, M.B. Marcus and P.A. Meyer. IV, 128 pages. 1986.

Vol. 1194: Complex Analysis and Algebraic Geometry. Proceedings, 1985. Edited by H. Grauert. VI, 235 pages. 1986.

Vol. 1195: J.M. Barbosa, A.G. Colares, Minimal Surfaces in $\mathbb{R}^3$. X, 124 pages. 1986.

Vol. 1196: E. Casas-Alvero, S. Xambó-Descamps, The Enumerative Theory of Conics after Halphen. IX, 130 pages. 1986.

Vol. 1197: Ring Theory. Proceedings, 1985. Edited by F.M.J. van Oystaeyen. V, 231 pages. 1986.

Vol. 1198: Séminaire d'Analyse, P. Lelong – P. Dolbeault – H. Skoda. Seminar 1983/84. X, 260 pages. 1986.

Vol. 1199: Analytic Theory of Continued Fractions II. Proceedings, 1985. Edited by W.J. Thron. VI, 299 pages. 1986.

Vol. 1200: V.D. Milman, G. Schechtman, Asymptotic Theory of Finite Dimensional Normed Spaces. With an Appendix by M. Gromov. VIII, 156 pages. 1986.

Vol. 1201: Curvature and Topology of Riemannian Manifolds. Proceedings, 1985. Edited by K. Shiohama, T. Sakai and T. Sunada. VII, 336 pages. 1986.

Vol. 1202: A. Dür, Möbius Functions, Incidence Algebras and Power Series Representations. XI, 134 pages. 1986.

Vol. 1203: Stochastic Processes and Their Applications. Proceedings, 1985. Edited by K. Itô and T. Hida. VI, 222 pages. 1986.

Vol. 1204: Séminaire de Probabilités XX, 1984/85. Proceedings. Edité par J. Azéma et M. Yor. V, 639 pages. 1986.

Vol. 1205: B.Z. Moroz, Analytic Arithmetic in Algebraic Number Fields. VII, 177 pages. 1986.

Vol. 1206: Probability and Analysis, Varenna (Como) 1985. Seminar. Edited by G. Letta and M. Pratelli. VIII, 280 pages. 1986.

Vol. 1207: P.H. Bérard, Spectral Geometry: Direct and Inverse Problems. With an Appendix by G. Besson. XIII, 272 pages. 1986.

Vol. 1208: S. Kaijser, J.W. Pelletier, Interpolation Functors and Duality. IV, 167 pages. 1986.

Vol. 1209: Differential Geometry, Peñíscola 1985. Proceedings. Edited by A.M. Naveira, A. Ferrández and F. Mascaró. VIII, 306 pages. 1986.

Vol. 1210: Probability Measures on Groups VIII. Proceedings, 1985. Edited by H. Heyer. X, 386 pages. 1986.

Vol. 1211: M.B. Sevryuk, Reversible Systems. V, 319 pages. 1986.

Vol. 1212: Stochastic Spatial Processes. Proceedings, 1984. Edited by P. Tautu. VIII, 311 pages. 1986.

Vol. 1213: L.G. Lewis, Jr., J.P. May, M. Steinberger, Equivariant Stable Homotopy Theory. IX, 538 pages. 1986.

Vol. 1214: Global Analysis – Studies and Applications II. Edited by Yu.G. Borisovich and Yu.E. Gliklikh. V, 275 pages. 1986.

Vol. 1215: Lectures in Probability and Statistics. Edited by G. del Pino and R. Rebolledo. V, 491 pages. 1986.

Vol. 1216: J. Kogan, Bifurcation of Extremals in Optimal Control. VIII, 106 pages. 1986.

Vol. 1217: Transformation Groups. Proceedings, 1985. Edited by S. Jackowski and K. Pawalowski. X, 396 pages. 1986.

Vol. 1218: Schrödinger Operators, Aarhus 1985. Seminar. Edited by E. Balslev. V, 222 pages. 1986.

Vol. 1219: R. Weissauer, Stabile Modulformen und Eisensteinreihen. III, 147 Seiten. 1986.

Vol. 1220: Séminaire d'Algèbre Paul Dubreil et Marie-Paule Malliavin. Proceedings, 1985. Edité par M.-P. Malliavin. IV, 200 pages. 1986.

Vol. 1221: Probability and Banach Spaces. Proceedings, 1985. Edited by J. Bastero and M. San Miguel. XI, 222 pages. 1986.

Vol. 1222: A. Katok, J.-M. Strelcyn, with the collaboration of F. Ledrappier and F. Przytycki, Invariant Manifolds, Entropy and Billiards; Smooth Maps with Singularities. VIII, 283 pages. 1986.

Vol. 1223: Differential Equations in Banach Spaces. Proceedings, 1985. Edited by A. Favini and E. Obrecht. VIII, 299 pages. 1986.

Vol. 1224: Nonlinear Diffusion Problems, Montecatini Terme 1985. Seminar. Edited by A. Fasano and M. Primicerio. VIII, 188 pages. 1986.

Vol. 1225: Inverse Problems, Montecatini Terme 1986. Seminar. Edited by G. Talenti. VIII, 204 pages. 1986.

Vol. 1226: A. Buium, Differential Function Fields and Moduli of Algebraic Varieties. IX, 146 pages. 1986.

Vol. 1227: H. Helson, The Spectral Theorem. VI, 104 pages. 1986.

Vol. 1228: Multigrid Methods II. Proceedings, 1985. Edited by W. Hackbusch and U. Trottenberg. VI, 336 pages. 1986.

Vol. 1229: O. Bratteli, Derivations, Dissipations and Group Actions on C*-algebras. IV, 277 pages. 1986.

Vol. 1230: Numerical Analysis. Proceedings, 1984. Edited by J.-P. Hennart. X, 234 pages. 1986.

Vol. 1231: E.-U. Gekeler, Drinfeld Modular Curves. XIV, 107 pages. 1986.

Vol. 1232: P.C. Schuur, Asymptotic Analysis of Soliton Problems. VIII, 180 pages. 1986.

Vol. 1233: Stability Problems for Stochastic Models. Proceedings, 1985. Edited by V.V. Kalashnikov, B. Penkov and V.M. Zolotarev. VI, 223 pages. 1986.

Vol. 1234: Combinatoire énumérative. Proceedings, 1985. Edité par G. Labelle et P. Leroux. XIV, 387 pages. 1986.

Vol. 1235: Séminaire de Théorie du Potentiel, Paris, No. 8. Directeurs: M. Brelot, G. Choquet et J. Deny. Rédacteurs: F. Hirsch et G. Mokobodzki. III, 209 pages. 1987.

Vol. 1236: Stochastic Partial Differential Equations and Applications. Proceedings, 1985. Edited by G. Da Prato and L. Tubaro. V, 257 pages. 1987.

Vol. 1237: Rational Approximation and its Applications in Mathematics and Physics. Proceedings, 1985. Edited by J. Gilewicz, M. Pindor and W. Siemaszko. XII, 350 pages. 1987.

Vol. 1238: M. Holz, K.-P. Podewski and K. Steffens, Injective Choice Functions. VI, 183 pages. 1987.

Vol. 1239: P. Vojta, Diophantine Approximations and Value Distribution Theory. X, 132 pages. 1987.

Vol. 1240: Number Theory, New York 1984–85. Seminar. Edited by D.V. Chudnovsky, G.V. Chudnovsky, H. Cohn and M.B. Nathanson. V, 324 pages. 1987.

Vol. 1241: L. Gårding, Singularities in Linear Wave Propagation. III, 125 pages. 1987.

Vol. 1242: Functional Analysis II, with Contributions by J. Hoffmann-Jørgensen et al. Edited by S. Kurepa, H. Kraljević and D. Butković. VII, 432 pages. 1987.

Vol. 1243: Non Commutative Harmonic Analysis and Lie Groups. Proceedings, 1985. Edited by J. Carmona, P. Delorme and M. Vergne. V, 309 pages. 1987.

Vol. 1244: W. Müller, Manifolds with Cusps of Rank One. XI, 158 pages. 1987.

Vol. 1245: S. Rallis, L-Functions and the Oscillator Representation. XVI, 239 pages. 1987.

Vol. 1246: Hodge Theory. Proceedings, 1985. Edited by E. Cattani, F. Guillén, A. Kaplan and F. Puerta. VII, 175 pages. 1987.

Vol. 1247: Séminaire de Probabilités XXI. Proceedings. Edité par J. Azéma, P.A. Meyer et M. Yor. IV, 579 pages. 1987.

Vol. 1248: Nonlinear Semigroups, Partial Differential Equations and Attractors. Proceedings, 1985. Edited by T.L. Gill and W.W. Zachary. IX, 185 pages. 1987.

Vol. 1249: I. van den Berg, Nonstandard Asymptotic Analysis. IX, 187 pages. 1987.

Vol. 1250: Stochastic Processes – Mathematics and Physics II. Proceedings 1985. Edited by S. Albeverio, Ph. Blanchard and L. Streit. VI, 359 pages. 1987.

Vol. 1251: Differential Geometric Methods in Mathematical Physics. Proceedings, 1985. Edited by P.L. García and A. Pérez-Rendón. VII, 300 pages. 1987.

Vol. 1252: T. Kaise, Représentations de Weil et $GL_2$ Algèbres de division et $GL_n$. VII, 203 pages. 1987.

Vol. 1253: J. Fischer, An Approach to the Selberg Trace Formula via the Selberg Zeta-Function. III, 184 pages. 1987.

Vol. 1254: S. Gelbart, I. Piatetski-Shapiro, S. Rallis. Explicit Constructions of Automorphic L-Functions. VI, 152 pages. 1987.

Vol. 1255: Differential Geometry and Differential Equations. Proceedings, 1985. Edited by C. Gu, M. Berger and R.L. Bryant. XII, 243 pages. 1987.

Vol. 1256: Pseudo-Differential Operators. Proceedings, 1986. Edited by H.O. Cordes, B. Gramsch and H. Widom. X, 479 pages. 1987.

Vol. 1257: X. Wang, On the C*-Algebras of Foliations in the Plane. V, 165 pages. 1987.

Vol. 1258: J. Weidmann, Spectral Theory of Ordinary Differential Operators. VI, 303 pages. 1987.

Vol. 1259: F. Cano Torres, Desingularization Strategies for Three-Dimensional Vector Fields. IX, 189 pages. 1987.

Vol. 1260: N.H. Pavel, Nonlinear Evolution Operators and Semigroups. VI, 285 pages. 1987.

Vol. 1261: H. Abels, Finite Presentability of S-Arithmetic Groups. Compact Presentability of Solvable Groups. VI, 178 pages. 1987.

Vol. 1262: E. Hlawka (Hrsg.), Zahlentheoretische Analysis II. Seminar, 1984–86. V, 158 Seiten. 1987.

Vol. 1263: V.L. Hansen (Ed.), Differential Geometry. Proceedings, 1985. XI, 288 pages. 1987.

Vol. 1264: Wu Wen-tsün, Rational Homotopy Type. VIII, 219 pages. 1987.

Vol. 1265: W. Van Assche, Asymptotics for Orthogonal Polynomials. VI, 201 pages. 1987.

Vol. 1266: F. Ghione, C. Peskine, E. Sernesi (Eds.), Space Curves. Proceedings, 1985. VI, 272 pages. 1987.

Vol. 1267: J. Lindenstrauss, V.D. Milman (Eds.), Geometrical Aspects of Functional Analysis. Seminar. VII, 212 pages. 1987.

Vol. 1268: S.G. Krantz (Ed.), Complex Analysis. Seminar, 1986. VII, 195 pages. 1987.

Vol. 1269: M. Shiota, Nash Manifolds. VI, 223 pages. 1987.

Vol. 1270: C. Carasso, P.-A. Raviart, D. Serre (Eds.), Nonlinear Hyperbolic Problems. Proceedings, 1986. XV, 341 pages. 1987.

Vol. 1271: A.M. Cohen, W.H. Hesselink, W.L.J. van der Kallen, J.R. Strooker (Eds.), Algebraic Groups Utrecht 1986. Proceedings. XII, 284 pages. 1987.

Vol. 1272: M.S. Livšic, L.L. Waksman, Commuting Nonselfadjoint Operators in Hilbert Space. III, 115 pages. 1987.

Vol. 1273: G.-M. Greuel, G. Trautmann (Eds.), Singularities, Representation of Algebras, and Vector Bundles. Proceedings, 1985. XIV, 383 pages. 1987.

Vol. 1274: N. C. Phillips, Equivariant K-Theory and Freeness of Group Actions on C*-Algebras. VIII, 371 pages. 1987.

Vol. 1275: C.A. Berenstein (Ed.), Complex Analysis I. Proceedings, 1985–86. XV, 331 pages. 1987.

Vol. 1276: C.A. Berenstein (Ed.), Complex Analysis II. Proceedings, 1985–86. IX, 320 pages. 1987.

Vol. 1277: C.A. Berenstein (Ed.), Complex Analysis III. Proceedings, 1985–86. X, 350 pages. 1987.

Vol. 1278: S.S. Koh (Ed.), Invariant Theory. Proceedings, 1985. V, 102 pages. 1987.

Vol. 1279: D. Ieşan, Saint-Venant's Problem. VIII, 162 Seiten. 1987.

Vol. 1280: E. Neher, Jordan Triple Systems by the Grid Approach. XII 193 pages. 1987.

Vol. 1281: O.H. Kegel, F. Menegazzo, G. Zacher (Eds.), Group Theory. Proceedings, 1986. VII, 179 pages. 1987.

Vol. 1282: D.E. Handelman, Positive Polynomials, Convex Integral Polytopes, and a Random Walk Problem. XI, 136 pages. 1987.

Vol. 1283: S. Mardešić, J. Segal (Eds.), Geometric Topology and Shape Theory. Proceedings, 1986. V, 261 pages. 1987.

Vol. 1284: B.H. Matzat, Konstruktive Galoistheorie. X, 286 pages. 1987.

Vol. 1285: I.W. Knowles, Y. Saitō (Eds.), Differential Equations and Mathematical Physics. Proceedings, 1986. XVI, 499 pages. 1987.

Vol. 1286: H.R. Miller, D.C. Ravenel (Eds.), Algebraic Topology Proceedings, 1986. VII, 341 pages. 1987.

Vol. 1287: E.B. Saff (Ed.), Approximation Theory, Tampa. Proceedings, 1985–1986. V, 228 pages. 1987.

Vol. 1288: Yu. L. Rodin, Generalized Analytic Functions on Riemann Surfaces. V, 128 pages, 1987.

Vol. 1289: Yu. I. Manin (Ed.), K-Theory, Arithmetic and Geometry. Seminar, 1984–1986. V, 399 pages. 1987.